the twilight of
world steam

the twilight of
world steam

ron ziel and mike eagleson

HAMLYN

London · New York · Sydney · Toronto

Racing through the junction at Salzbergen, Germany, an 012 Pacific tops 60 m.p.h. in December 1970. *Ron Ziel*

Other Books by the Authors
SOUTHERN STEAM SPECIALS

Books by Ron Ziel
*THE TWILIGHT OF STEAM LOCOMOTIVES
THE STORY OF STEAMTOWN AND EDAVILLE
STEEL RAILS TO VICTORY*

Books by Ron Ziel and George H. Foster
*STEEL RAILS TO THE SUNRISE
STEAM IN THE SIXTIES*

A MADISON SQUARE PRESS BOOK®

Originally published in the U.S.A. by
Grosset and Dunlap Publishers, New York.

This edition published in 1975 by
The Hamlyn Publishing Group Limited
London · New York · Sydney · Toronto
Astronaut House, Feltham, Middlesex, England

Copyright © Ron Ziel and Mike Eagleson 1973

ISBN 0 600 38707 0

Printed by Litografia A. Romero, S.A.
Santa Cruz de Tenerife, Canary Islands (Spain)

D. L. TF. 314 - 1975

Spinning driving wheels and "monkey motion" of an Argentinian 2-10-2 working the yard at General Martin Miĝuel de Guemes. *Ron Ziel*

Erupting geysers of steam, a 2-8-4 wheels an express out of the main station in Sibiu, Rumania.

Ron Ziel

Contents

Dedication

To my mother, Mrs. Edith Ziel Brandewie, who encouraged me at an early age to partake of the steam legend by taking me down to the tracks to watch the trains, and who never failed in her support (morally and at times financially) even to the extent of leaving her warm Florida home in several mid-winters to come north and be caretaker of my house and step-mother to my cats, while I gallivanted 'round the world in search of steam.

-Ron Ziel

To my loving wife, Naomi, who sacrifices her own pleasures, normal household comforts and even bodily safety to pursue with me, on six continents, my fanatical obsession for the steam locomotive. She consoles when rain clouds darken, is patient through endless trackside waits, encourages while motorcading on treacherous roads and has suffered jail with me in totalitarian lands — only to return home an impoverished darkroom widow.

-Mike Eagleson

Acknowledgments

The authors are deeply indebted to many individuals, organizations and government officials in many countries for giving of their time and their knowledge, as well as their connections in high places, to assist in the completion of this work. Some offered assistance which was truly outstanding, including the results of their own photographic talents. They include: Victor Hand, L.G. Marshall, Paul S. Stephanus, Howard Serig, Lance King and the *Continental Railway Journal*, Frank Stenvall and *Railway Scene*, Badger Roberts, Cdr. William D. Middleton, U.S. Navy, Roy Christian, John Briggs, Gordon Roth, John Dziobko, Brad S. Miller of Mobile Fidelity Records and Edward L. Conklin, III. Family members who assisted and at times sacrificed, were: Mrs. Jean Egan Eagleson, Naomi Eagleson and Mrs. Edith Ziel Brandewie. Special mention for assistance of inestimable value in several countries is reserved for: Reino Kalliomäki in Finland, Rüdiger Weber in Germany, Etienne Mouradian, Treasurer of the Lebanese State Railway, Dr. Gilberto C. Cabral, Chief Mechanical Engineer of the *Teresa Cristina* in Brazil, William Eduardo Fredes, chief photographer of *Ferrocarriles Argentinos* and Mrs. Stefania Dabkowska of the Ministry of Communication in Poland. Travel plans were expedited by Leroy Beaujon and by the Osborne Travel Bureau of Southampton, New York, and John Osborne and Adele Lamb. Various American embassies of the United States Department of State were also very helpful in getting permission for the authors to take pictures and in retrieving confiscated film. Finally, many offices of automobile rental agencies, especially those of the Hertz Company, went way "beyond the call of duty" in assistance, ranging from locating rare locomotives to securing hotel rooms.

Introduction

North American steam enthusiasts were both the luckiest and the unluckiest. The railroads of North America built and operated the largest and most awesome steam locomotives the world has ever seen, but North American railroads were also the first to succumb to the diesel locomotive. The economics of internal combustion made the quick elimination of steam a certainty and by 1960 the last fires had been dropped on the mainlines, except for occasional excursion trains.

Some railway enthusiasts found that the excitement of steam was not the only thing that had drawn them to trackside. The timeless spectacle of transportation was still interesting even with look-alike diesels at the head of every train. Some railfans even found that they could get excited about the new locomotives. Others could not; steam disappeared and they turned to other interests.

When the last smoke plume drifted skyward, many North American rail enthusiasts turned their attention overseas. Fast transport and a rising standard of living made it possible for railfanning to take on an international dimension. First it was Canada, Mexico and Central America, where steam operations lasted a few years longer than at home. Then the introduction of trans-Atlantic jet travel made Europe easily accessible and more and more Americans went to look at the graceful steamers of British Railways, their mechanically sophisticated cousins in France and the utilitarian machines of the *Deutsche Bundesbahn*.

But the diesels were making inroads all over the world. The intrepid steam fan attempted to stay one step ahead of the salesmen from General Motors and English Electric. The search spread to Africa, Australia, South America and Asia.

The North American steam locomotive had been exported to many parts of the world and in those countries where engines from Baldwin, American and the Montreal Locomotive Works were still running, the North American felt at home. The British, German, French and Japanese locomotive builders had also exported locomotives by the thousands, and the influences of the various schools of design had spread over the railway world like a patchwork quilt. To a North American eye, many of the foreign designs were hard to accept, but the universal excitement of steam, steel and loud stack talk came through in any language.

Often, the last steam operations in a particular country are in remote areas seldom visited by outsiders and the visiting photographer must overcome problems of transport, language, food supply, and accommodation that would have made Dr. Living-stone think twice before leaving for Africa. The authors of *The Twilight of World Steam* have travelled to most of the fifty-five countries represented in the book to photograph trains and have drawn upon the collections of other hard-travelling photographers to fill in some of the countries which they could not visit themselves. The pictures tell the story of steam in its prime and of steam on its last legs.

Perhaps India will be the last country to operate steam; perhaps China. Just as twilight has already fallen on the steam locomotive in North America and many parts of Europe, within a few years it will most certainly fall on the rest of the world.

–Victor Hand

Mike Eagleson

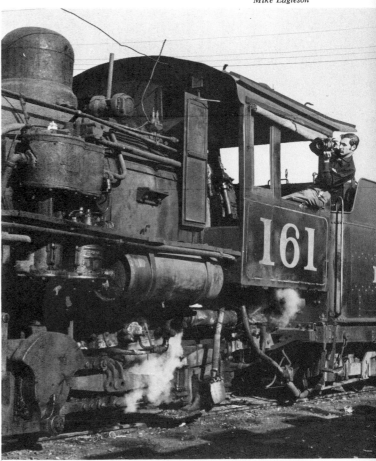

VICTOR HAND is perhaps the most renowned and certainly the most traveled of the young generation of international photographers. He has a New York State Law Degree and has made his career in railroad management. He travels abroad as much as time permits and has contributed to numerous railway books and periodicals.

Foreword

For six generations, the destiny of America and especially of her young boys was irrevocably coupled to the romantic pageant of the steady parade of main-line steam power that marched incessantly through virtually every sizable hamlet and city. The authors were at once fortunate and unfortunate to have been of the last generation to witness the great sooty spectacle. They were fortunate to know the era when every starry-eyed boy dreamed of taking the throttle of the great iron steed at the headend of a transcontinental limited. But they were also unfortunate because the era of steam ended abruptly, almost as soon as they had begun their life-long love affair with it.

Through young manhood they pursued the ever-dwindling ranks of active shortline engines, and the steam excursion trains. In final desperation, they even ventured onto the properties of the gun-slingin', popcorn-peddling tourist railroads. With acute frustration becoming a way of life, could insanity be close behind? It could — and was.

Although regarded with a benign contempt by family, friends and business associates, the authors, in separately reached decisions, joined the growing ranks of international steam photographers and headed first to Canada and Mexico, then to western Europe, at various times in the 1960's and early '70's. Except for language, few problems were encountered in those early years, but with expanded itineraries due largely to the commissioning of this book — including eastern Europe, Africa, Asia, Australia and South America — the dangers of hostile governments, backwood natives, rugged terrain, disease, wild animals and thirty-year-old DC-3's became constant adversities.

Always just a jump ahead of the diesel salesmen or the electrification line crews, the international steam photographers had little say in planning their own trips. It was always the urgent question of what would be scrapped next and, if several countries were dieselizing at once as was usually the case, how to make the agonizing decisions as to what to see first. At times, imminent electrification spelled a brief reprieve for steam, since railway managements were reluctant to buy diesels which may be rendered redundant by catenary in a year or two.

The casual observer, including perhaps most readers of this monologue, has no concept of the misery, the privation, the scuzziness and the sheer inhumanity of some of the places endured by the authors. Trips to photograph foreign steam invariably begin in the office of one's employer, who refuses to grant repeated requests for six weeks off several times a year. The planning, scheduling of transport, raising funds and the last-minute reports of dieselization disasters, all tend to wreak havoc with the intricately planned itineraries. Upon returning home to hunt for a new job, scarce remaining funds prompt the American Express computer to excrete threatening demands.

With the west European countries rapidly phasing out steam, and the countries of eastern Europe having been more or less covered — usually without the consent of the police — attention was turned to the Near East and the Southern Hemisphere, where such civilized countries as the Republic of South Africa, Rhodesia, Australia and the larger South American nations are few and far apart. More often, the rule is filthy hovels masquerading as hotels, restaurants that smell like stables, hundreds of kilometers of unmarked, incredibly treacherous dirt roads and natives who do not even know the name of the next village, but who are fully cognizant of the fact that the railway enthusiast's camera represents two years wages working in the insect infested rice paddies.

Other occupational hazards endured by the authors and the contributors to this work include facing the wrong end of Soviet-made submachine guns, lice-covered blankets in village jails, taxi drivers who go to the wrong airports and poisonous snakes which live behind hotel room mirrors! After all of these vicissitudes are miraculously survived, some local *policia*, impressed with their own authority, may casually confiscate all of the film exposed during several harrowing weeks, for railway photography in many countries is regarded as sheer espionage!

During the eight years (1966 to 1973) covered by

The clock on the shop building of the Kingmoor Engine Shed in Carlisle, Scotland, reads 11:20 as the nighttime mist descends on the fabled moors, enveloping steam locomotives whose rest is troubled by nightmares of the coming of the diesels just a few months away.

Mike Eagleson

this dissertation, perhaps seventy nations still operated at least a modicum of steam power. Fifty-five are represented here; the remainder were unobtainable for various reasons. The People's Republic of China, perhaps the most important of all, had not yet been opened to the authors, while Cuba, North Vietnam and North Korea also were off-limits by mutual consent of both their governments and the United States. Syria, Uganda, the Congo and Ghana were considered too dangerous to attempt, while the Philippines, Angola, Uruguay, Paraguay and others were either marginal in their steam operations or unreachable by the authors. Also, it must be realized that it is impossible to present, in the limited space available, all of the great variety of locomotives which were seen and photographed during our travels. Literally hundreds of classes survived into the twilight of world steam. Much as the authors would have liked to, we could not include everything.

If the immediate past has been grueling and gruesome, at least it offered a relative abundance of steam power. But what of the future of world steam? A few countries of Latin America and western Europe will see steam activity through the mid-1970's. In eastern Europe, only Poland appears destined to offer some action beyond 1980. The Indian sub-continent, South Africa and China will be the last major steam countries to go, but only India is expected to last into the 1990's and, possibly, still dispatch some steam trains in the twenty-first century.

Both authors are well aware of the political, economic and racial policies of the greatly diverse spectrum of nations represented in this volume. The coverage of locomotives on their railways is by no means to be construed as acceptance of any of those policies. As historians recording the end of an era, the authors were concerned with that subject only. However, having spent considerable time in republics, democracies, socialist countries, communist countries, colonies and military and fascist dictatorships in pursuit of steam, they are fully cognizant of life — good and bad — under all forms of government.

The typical American railway enthusiast, having been deprived of steam in regular service for lo these many years, has developed a narrow, provincial attitude toward "those homely foreign engines, whose bald front ends are without pilots or headlights and are disfigured by buffers . . . ." Yet, what would appear more at home on American rails than the Baldwin locomotives of Brazil's *Teresa Cristina*, the big Vulcan Decapods in Turkey, the War Department 2-8-2's of Thailand, the Soviet 4-8-4's on the Trans-Siberian line and the many other operations shown here? For even the casual steam buff, the authors hope that this book will prove to be *the* definitive work on the twilight of world steam and that their time, tribulations and more than $30,000 in traveling expenses alone will have been worth the effort.

-Ron Ziel
-Mike Eagleson

March 25, 1973

Great Britain

Great Britain, where steam harnessed to steel first turned flanged wheels to do the useful work of man, dispatched her final steam locomotives in 1968 — a century and a half after it all began. A mere two years previously, 2,000 steam engines still had powered many of the express and local passenger trains, as well as handled goods moves and the shunting chores of many stations. The last generations of British steam were the distinctive creations of the four major railways which were consolidated into the British Railways system in 1948. Generally, the classes that had been the rosters of the Southern Railway, the Great Western, the London, Midland & Scottish, and the London & Northeastern, remained with the lines of their former owners. The final classes — designed and erected in the 1950's, after nationalization — were uncomplicated, standardized two-cylinder machines with interchangeable parts and were quite handsome.

Racing an express hard by the Scottish coast of the North Sea (*above*) one of the last A-2 Pacifics built for the old L.N.E.R., *Blue Peter*, appropriately named after the 1939 Derby winner, exhibits her thoroughbred lineage bound for Glasgow just five months before her withdrawal from service. *Blue Peter* has been preserved, hopefully to run again. Much of the mystique and fascination of Britain's railways in steam days was concealed from casual observers by the soot-encrusted brick walls of the locomotive sheds. One of the larger sheds, at York, contained running tracks, two turntables and a complete shop — all beneath one roof! From left to right (*left*) two B-1 Ten wheelers, a V-2 Prairie, and a World War II Austerity Consolidation. The newest of the "Merchant Navy" class 4-6-2's, introduced in 1941, No. 35030, *Elder Dempster Lines*, carries a Weymouth-bound express (*below*) through Brockenhurst, on rails of the former Southern mainline.

All photos this chapter, Mike Eagleson, Summer 1966.

Typical of British streamlining practice, the most famous of English passenger express engines were the A-4 Pacifics (*above*), the masterpiece of Britain's foremost locomotive designer, Sir Nigel Gresley, as exemplified by *Kingfisher*, passing Coupar Angus, Scotland, on her native L.N.E.R. Less renowned, but more numerous (140 engines in three name classes) were Southern Railway's identical "Battle of Britain," "West Country" and the larger "Merchant Navy." All had the unusual tapered design, personified by

Spitfire (*below*), named after the fighter aircraft that won the Battle of Britain. Between 1934 and 1947, 842 locomotives of the highly successful Stanier "Black Five" 4-6-0's were built for the L.M.S. No. 44768 (*upper right*), climbing Shap Incline, works hard on this difficult stretch of the London & North Western. Introduced in 1934, the Jubilee class Ten wheelers were built to a total of 191 within two Depression years. A Jubilee named *Ulster* pulls a parcels train (*lower right*) from the towered Wakefield station.

Despite the strong trend toward utilizing the British Standard types in the final days of steam, a few untypical inhabitants at the engine sheds lingered on. Gothic in appearance, 0-6-0T No. 47471 (*above*) rests with a sister tanker in Carlisle, Scotland's famed Kingmoor Engine Shed after a rugged day of shunting goods wagons. A Stanier-designed 2-8-0 (*above, right*) formerly on the books of the L.M.S., charges through Horbury-Ossett with a crowded summer passenger train. High-drivered 4-6-0 B-1 No. 61261 (*center, right*) prepares to depart Glasgow's Eastfield Shed for a tour of duty on a local parcels train. The Ferryhill Locomotive Depot in Aberdeen, Scotland (*below, right*) holds such treasures as A-4 Pacifics *Union of South Africa* and *Bittern*, under repair for limited service on Glasgow expresses. The relentless steam laborers such as the lad who cleaned out the cinders from the smokebox of War Department Austerity 2-8-0 No. 90617 (*left*) at Normanton, will be long remembered in the annals of British railway history. Such mechanics had to be familiar with the workings of all classes and how to service them — right up until their final calls.

Had steam development in Britain continued, utilizing the vast reserves of local coal rather than having to depend on potentially hostile overseas countries for diesel oil, the last generation of steam engines — which were the only ones developed after nationalization — were a good indication of "what might have been." Mindful of the operating characteristics of the separate former independent companies and working with their engineering staffs, British Railways accepted new designs in five wheel arrangements and twelve sub-classes. This ambitious program was launched in 1951 with the construction of the "Britannia" class Pacifics, such as *Lord Rowallen* shown cresting Beattock Summit (*upper right*) with the aid of another standard — a 2-6-4T — banking on the rear. A 2-6-4 (*above*) about to leave London's Waterloo Station with a local. In addition to 200 Moguls built after 1952, there were 252 Ten wheelers, such as No. 75077 (*below*) racing through Basingstoke, and about 250 Decapods, such as No. 92165 (*lower right*) pulling a mixed goods train upgrade from Horbury-Ossett. When the sixty-six standard Pacifics are totaled in, nearly 1,000 ultra-modern steam locomotives had been erected between 1951 and 1960 — all being withdrawn by 1968. Dieselization of the birthplace of steam railroading was accomplished in record time — but at what an insidious price to the national soul of Britannia!

Portugal

The Portuguese railway system (*Companhia do Caminhos de Ferro Portugueses*) was a living museum in the last years of steam, with most of its power having been built prior to World War I and all of it the most immaculately maintained in Europe. Although only about one hundred locomotives remained on the combined rosters of the broad and narrow gauge lines into the mid-1970's, these were highly concentrated at several terminals in the northern part of the country and repeated reports of their demise were invariably mistaken.

The mountains and river valleys of northern Portugal offered ample opportunity for good photography of both gauges of steam power. The main 5′ 6″ gauge line from Salamanca, Spain, to Porto, on the Atlantic seacoast of Portugal, runs along the Rio Douro. Four narrow gauge lines branch off the main line and wind their way up into the north mountains, toward the Spanish border. At both Régua and Tua, huge steel spans on masonry abutments cross tributaries feeding the Rio Douro. As the morning fog lifts over the hills, a 1923 Henschel 2-4-6-0T Mallet, No. E-214 (*left*) brings a morning mixed train off the Corgo Valley line and into Régua across the dual gauge bridge virtually over the fork of the Douro and Corgo

Rivers. Medium-distance passenger services were handled by a class of twenty-eight handsomely appointed 2-6-4 Baltic tank engines, built in Switzerland and Germany from 1916 to 1929, with an additional one homemade in Lisbon in 1944. Most survived into the last years of steam, including No. 070 — the lowest numbered, but the newest built — wheeling a string of old wooden coaches (*below*) past less fortunate sisters awaiting scrap at Contumil in June 1968. Bound for Porto from Régua in the background in January 1971, a Henschel 2-6-4T from 1929 (*above*) blasts out of one of the many tunnels on the vineyard-studded Douro Valley line.

Two photos, Ron Ziel   Below, Mike Eagleson

Of the fourteen different classes of narrow gauge locomotives, encompassing seven wheel arrangements, the
smallest and obviously the "mascot" of the system was a
Henschel 0-4-0T. Built in 1922, the 0-4-0T was used
alternately with an older and larger 0-6-0T to switch the
dual gauge yard at Régua. So beloved was this diminutive shunter (*left*) that the fireman actually spent much
of his lunch hours wiping every surface and reshining all
the brass. Once, while switching a filthy oil tank car,
some black goo chanced to alight on the gleaming red
pilot beam and its radiant raised brass numerals. All
shunting ceased for five minutes as the fireman wiped
each and every spatter of the offending glop from his
precious charge.

Ron Ziel

One of twelve meter gauge 0-4-4-0T's which were in
commuter service out of Trinidade Station in Porto,
No. E-169, shown in June 1971 (*right*) along with her
sisters, built by Henschel in 1905 and 1908, was still very
active well into the 1970's. It was incongruous indeed to
view the rush hour mob scene as dapper businessmen
and mini-skirted office girls barged aboard the rickety
ancient wooden coaches after emerging from their ultra-
modern office buildings near the terminal. Puttering
through the Twentieth Century on board the Nine-
teenth!

L. G. Marshall

Mike Eagleson

Although the largest locomotives as the end of steam
neared in Portugal, the eighteen broad gauge Henschel
Pacifics, exemplified by No. 558 (*left*) at Contumil
engine depot in June 1968, apparently were retired three
years later. But examples of fully withdrawn classes on
the Iberian peninsula frequently are resurrected from the
ranks of condemned locomotives. The first of these
4-6-2's, No. 501, was revived in 1971, making her the
last to operate. Just a few months later, however, she
suffered heavy damage in a right-of-way dispute with an
electric locomotive. That mishap does not preclude another Pacific being returned to service.

Mike Eagleson

Portugal's largest tank engine class was the hefty
Henschel broad gauge 2-8-4T of the mid-1920's. Ten
were built, including No. 0188 (*right*) shown at Porto in
1968. Locomotives without tenders, which carried their
fuel and water on the same frame as the boiler, rarely
were developed beyond the shunting stage in North
America — usually as 0-4-0T or 0-6-0T types. Many
European railways went much further in their tank designs, building far-ranging road engines capable of great
speeds and with formidable tractive effort ratings. Wheel
arrangements of 2-8-2T, 2-8-4T, 4-8-4T, 2-10-2T, and
even 2-12-4T proved how far the builders could go with
the tenderless principle of a locomotive carrying its
sustenance on its back and still create a handsome
machine.

20

Mike Eagleson

The locomotives of Portugal could really put on a show when the circumstances of gradiant, tonnage, damp mountain air and perhaps a novice fireman combine to create conditions such as those under which 4-6-0 No. 284 (*left*) departs Régua with a mainline freight on January 15, 1971. Lending credence to the museum atmosphere of the rail system were such engines as 0-4-0T No. 003 (*above*), the Contumil shop switcher, which was built in 1890. Shortly after this photo was made in 1968, she was replaced by a similar machine — built in 1881! The turntable at Régua (*below*), here moving meter gauge 2-4-6-0T No. E-202, was another splendid anachronism, with its man-powered chain gear drive.

Two photos, Ron Ziel

One of the attractions that so endeared railway photographers to Régua was the little parish church just beyond the white wall that enclosed the engine terminal. It was reverently known as the "Cathedral of Steam" to the many enthusiasts who came to see God's little engines in this portion of the Rio Douro Valley. Pacific No. 553, Ten wheeler No. 241, and narrow gauge compound Mallet No. E-204 (*left*) await their next calls in the summer of 1968. Two-and-a-half years later, 2-6-4T No. 093 (*below*) stomps onto the turntable while 2-8-0 No. 712 and inside cylinder 4-6-0 No. 286 observe the antics of the tanker on the merry-go-round. Although very French in appearance, 4-6-0 No. 238 (*above*) was a four-cylinder compound built by the Berlin firm of Borsig in 1904. As she passes behind meter gauge 0-6-0T No. E-54, a rare outside Stephenson gear shunter built by Esslingen in 1889, she presents a scene of contrast that occurred a score of times daily in the Régua yard. *Overleaf, top:* northbound on the Corgo line, a Henschel Mallet works upgrade through Alvacoes, stating the case for German-built Portuguese steam power.

Facing page, Mike Eagleson   This page, Ron Ziel

Overleaf, top, Ron Ziel

Spain

The locomotives of the 5′ 6″ broad gauge lines of Spain will be remembered for their immensity; indeed in three major classes, engine weights exceeded 200 tons in working order — the only steam power in Europe to do so. The charms of Spain, which have endeared this ancient land to so many, thoroughly permeated the rosters of its steam giants. Almost to the end, little centenarian shunters worked sheds that housed sisters in steam built 104 years later. The tribulations and sorrows of Spain proved a mixed blessing to the R.E.N.F.E. (*Red Nacional de los Ferrocarrilles Españoles*) system, for the Civil War of the late 1930's had devastated the railways, and World War II delayed modernization — prolonging *la era de Vapor* for many years. The diminutive 0-6-0's (some dating back to 1857), the huge tonnage haulers and the Garratts (the final batch of European steam locomotives, built in 1961) worked together beneath the vast El Greco skies and fell silent — the last departing in 1973, carrying with them forever as their freight a part of that intangible legacy that is the Hispanic mystique.

Resplendent in green livery, No. 242F.2010 (*left*), one of only ten magnificent 4-8-4's built in 1955 and 1956 by Maquinista of Barcelona, passes the village church at Pancorbo, beneath the newly installed catenary which, a few weeks later in that summer of 1968, forced these engines off the exalted expresses and into secondary services. These 4-8-4's, the only green R.E.N.F.E. engines, along with fifty-seven heavy 4-8-2's and twenty 2-10-2's, were the only locomotives in Europe to exceed 200 tons in working order. Some continued to receive heavy repairs into the early 1970's and were among the last steamers in regular service. Crossing the Rio Segre at Lerida (*above*), a pair of 4-8-0's are silhouetted against the early morning sky as they move a troop extra on a June morning in 1968.

All photos this chapter, Mike Eagleson (except as noted)

Above, Ron Ziel

Spain's standard, modern, general purpose steam power was embodied in 242 big, handsome Mikados, erected between 1952 and 1960. The first series of twenty-five, built by North British, included No. 141F.2112 (*upper left*), shown working her single Kylala blast pipe as she works the hill beneath the main highway overpass near Noáin in January 1971. The remaining 217, constructed by the four major Spanish locomotive factories, were characterized by double chimneys and separated steam dome and sandbox, as typified by the trim and clean 141F.2340 (*left*) climbing the grade at Aldeseca in 1968. While a section crew replaces worn rails around the turntable at Lerida (*below*), 2-8-2 No. 141F.2417 rides the table; a 4-8-4T and a 4-8-0 await their turns behind. Sporting huge teutonic-style smoke lifters and rolling on 75-inch spoked driving wheels, 4-8-4 No. 242F.2007 (*above*) wheels a Madrid-bound express through Castil de Peones.

The most common type of all the road power classes, with 322 examples, was the 4-8-0 wheel arrangement used for virtually any assignment except heavy passenger service. Together with the Mikados, the Twelve wheelers wrote the final chapter of the Spanish steam story in the early 1970's. In June 1968, one of the older (1927) 4-8-0's, No. 240F.2309 (*below*), double-heads up the steep grade at Baños de Montemayor, assisted by 2-8-2 No. 141F.2388. An even older sister, 240F.2287 (*upper right*), wheels a local peddler freight near Rio Tajo. The bigger and final 4-8-0 development, characterized by a 1942 example, No. 240F.2528, emerges from the tunnel at San Felices (*center right*). The largest, heaviest, and most powerful steam locomotive ever to run on the European continent was the R.E.N.F.E. 2-10-2. No. 151F.3102 (*below, right*) takes a turn on the table at Arcos de Jalon.

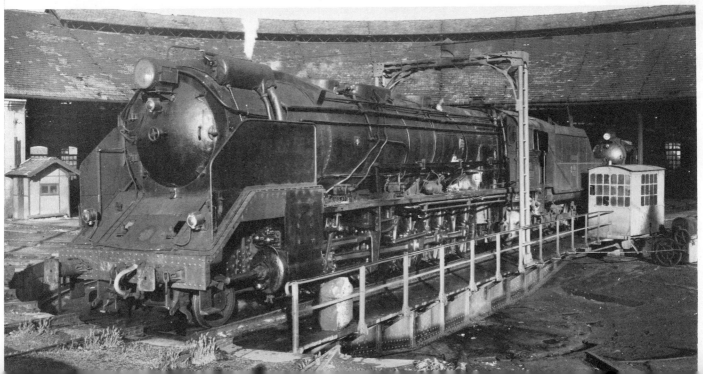

In 1939, sensing the worldwide trend toward stream-lining locomotives, one of the predecessor lines of R.E.N.F.E. ordered ten 4-8-2's of a semi-streamlined configuration. They saw hard service for the ensuing three decades. No. 241F.2103 (*above*) pulled a heavy freight through Pancorbo in her last year of service. For three consecutive days, photographers Eagleson and Victor Hand defied a six-foot black snake (apparently a reptilian railway enthusiast) who lived in sooty recluse beneath the brick arch bridge (*left*) here framing 4-8-2 No. 241F.2090 at Tejares. Passing beneath a ninety-seven-arch aqueduct built in 1790 to carry water to Pamplona, No. 241F.2247 (*upper right*), one of the last active R.E.N.F.E. Mountain types, crosses the Elorz Valley near Noáin on January 22, 1971. A brigade of 4-8-2's and two 2-10-2's (*right*) wait out the Spanish summer night at Zaragoza. Crossing the Tormes River bridge at Salamanca, one of the earliest series 4-8-2's, No. 241F.2077 (*below*), built in the 1920's, reflects the evening sunlight in June 1968.

Above, Ron Ziel

Perhaps the most distinctive locomotive ever designed anywhere was the highly unorthodox Beyer-Garratt. The boiler was slung between two engines for maximum tractive effort with minimum weight on axles. Sort of an ultimate tank locomotive, the fuel and water bunkers were supported by the engine wheels and had no separate trucks. Developed by the famous British firm of Beyer-Peacock, the Garratts were erected in most popular wheel arrangements. Both engines always had the same layout, but were applied back to back (i.e., 4-6-2 + 2-6-4, 2-6-0 + 0-6-2, etc.). These locomotives never ran "backwards" since the cylinders at each end always were up at the buffer beam. Although several European countries experimented with Garratts, only England and Spain retained them in regular service. The Spanish ones, in 2-8-2 and 4-6-2 types, were by far the best known and the longest-lived in Europe. A 2-8-2 Garratt (*opposite, above*) leaves the station at Picamoixons while a sister waits to follow with a freight. Another 2-8-2 Garratt, No. 0427, is shown in night action (*opposite, below*) at Lerida. Of the fifteen 2-8-2 Garratts, the ten built in 1960–'61 were the last new steamers erected for domestic use in Europe. All were retired by the end of the decade, but two 2-8-2 and one 4-6-2 Garratts were recalled to active duty for a few weeks early in 1971. Since Spain possesses little coal reserves — all of it poor grade — few coal-fired locomotives survived beyond the time it took to convert them to oil. Two exceptions are shown here in 1968. Ten wheeler No. 230.2090 (*upper left*), among the last survivors of sixty built in Germany in 1909, leaves Lerida at twilight with a local passenger train. Using low-grade coal compressed into briquettes, an ancient 0-6-0, No. 030.2564, the last assigned to a passenger run, departs Fuente de San Esteban (*below*) bound for the Portuguese border. One of the final examples of the ninety 4-8-4T's built in Spain in the mid 1920's, No. 242.0286 leaves Lerida (*upper right*) with an evening local. From 2,700 active steam locomotives to complete withdrawal in just seven years — an unenviable record and a tragic finale to one of the most beautiful of the world's steam operations.

France

A 144-year tradition of innovation and bold experimentation in the development of highly sophisticated steam locomotion received the *coup de grâce* in the summer of 1972, when the last of the simple American-built Liberation class Mikados were retired. Prior to that ironic finale, the designers of French steam power — true masters in their field — had long led the world in locomotive development. To those who maintained an intimate relationship with the locomotives of the *Société Nationale des Chemins de Fer Français* (S.N.C.F), memories of these thoroughbreds from the stable of world steam will always recall high-spoked driving wheels partially caressed by low running boards and splashers in the Anglican manner, propelled by steam expansion through the intricacies of four-cylinder compounding. Mainline passenger power often presented an illusion of running at 90 m.p.h. even when stationary, with rakishly angular smoke deflectors, canted cylinders, dainty-appearing rods and valve gear and swept-back cab lines.

The epitome of steam power on the S.N.C.F. was embodied in the final express passenger class: the 241P, thirty-five of which were built from 1948 to 1952. They were compounds but could be operated high pressure on all four cylinders at low speed, adding much needed power to get heavy trains under way. In January 1968, just a year and a half before her last run, No. 241.P 22 (*right*) works as she was designed to do — pulling an express upgrade at speed, bound for Nevers from Saincaize Junction. Like so many high-stepping passenger ladies before her, No. 241.P 16 (*lower right*) was humiliated in her final months when she was assigned to local freights out of Nantes. Still riding high at the time, many of the American-built 141.R class, such as No. 607 (*below*) shown wide open while working a mainline freight southward through Meauce, were destined to write the final chapter in the twilight of French steam, just four years later.

All photos this chapter, Mike Eagleson (except as noted)

Gallic nocturne. In rain or on dry cinders, at rest or in rapid dispatch or being fondled by their attentive groomers, steam locomotives of the S.N.C.F. acquire the mystic aura common to all of their kind, throughout the world, when stalked by night. Mikado No. 141.C 182 (*upper left*) is at Nantes; commuter tank engine No. 141.TC 59 (*lower left*) at Gare du Nord, Paris; 141.R 473 (*above*) gets her ashes dumped at Calais. Pacific No. 231.G 645 (*right*) blasts out of Nantes while (*below*) 241.P 12 is at LeMans, 12 minutes after midnight, on January 7, 1968.

Heavy shunting power in the form of 0-10-0T No. 050.TQ 33 (*upper left*) and Pacifics such as No. 231.K 31 (*above*) could still be seen in action virtually side by side as late as 1968. The World War I-era 4-6-2 was assigned to the *Flèche d'Or* (*Golden Arrow*) — the last named train on the S.N.C.F. to roll behind steam. No sooner had this express passed, when the author turned 90 degrees in place, flipped his film holder, and got the 0-10-0T. All of this commotion occurred at Pont de Briques. Yard scenes at Montluçon on January 13, 1968: 2-8-2T No. 141.TA 416, another World War I veteran (*left*) shares the roundhouse tracks with a half dozen 141.E's and F's. Eight-wheel switcher No. 040.TX 9 (*lower left*) was one of sixty-seven such machines built to German specifications from 1944 to 1947. The two examples of the 680 141.E's (*right and below*) built in the decade and a half following World War I and rebuilt in the 1940's, were simple in appearance but typically French in lineage.

When France was liberated from German occupation in 1944, only 3,000 of the 17,000 prewar steam engines of the S.N.C.F. remained operable. France needed a rapid infusion of simple, rugged, mixed-traffic locomotives and turned to North America where the five major U.S. and Canadian builders erected 1,340 two-cylinder 2-8-2 Mikados in less than two years, 1945–1947. The 141.R Liberation engines, although less efficient than the more sophisticated native classes, easily made up any shortcomings by their availability and low maintenance needs. By 1951, when they comprised just 15 percent of the active S.N.C.F. roster, the 141.R's ran double the average daily mileage of the French classes, totaling nearly 30 percent of all engine mileage and hauled 45 percent of the total tonnage, not to mention passenger operations. By January 1971, when fewer than one hundred of the French-built locomotives remained in service, more than four hundred 141.R's still worked around the country. 141.R 575 (*left*) doubleheaded a diesel freight at La Celle at that time and the coal dock at Nevers (*above*) still fed a steady parade of American 2-8-2's, such as 141.R's 409 and 235. The last doubleheading of 141.R's was done near Lake Geneva, with trains such as the freight (*below*) bound for Annemasse behind oil-burning Nos. 1019 and 921. For all of the well-earned accolades bestowed upon the superb steam power designed and built in France, it fell upon these spartan, mass-produced, simple, and typically American locomotives to suffer the final humiliation — and to know the bittersweet glory — of closing out *l'ère de Vapeur*.

Three photos, Ron Ziel

Italy

In the mid-1960's, Italy, with fewer than 1,000 steam locomotives remaining on the roster (only half in service at any time, with the remainder stored or awaiting repairs), was far from being considered among the more steamy of Western European nations. By the mid-1970's, however, with England, France, and Spain having phased out steam completely and with Germany, Portugal, and Greece doing the same, the "Boot" had become the second most steam-populated country in Western Europe. The Italian State Railway, *Ferrovie dello Stato* (F.S.), had been much slower in its motive power replacement program, and more than 800 steam engines are still carried on the active lists, with about 300 operating on any given day. Few locomotive connoisseurs have ever accused F.S. engines of being handsome or impressive. They are generally very small, bald looking, painted a flat black, and some classes look downright hideous. To many observers, however, these are most desirable traits in a locomotive fleet. These foregoing conditions are brought on by the fact that the Italians apparently are experts at prolonging the lives of steam locomotives. Unlike the other countries, no new Italian engines for domestic use have been built since the 1930's, and the heaviest machines surviving to the last years of steam were rather light 2-8-0 Consolidations.

The Italians did build a formidable class of large 2-10-2's in the mid-1950's — the first steam engine construction in twenty years — but all twenty of the big-boilered behemoths were on order by the Hellenic State Railways of Greece. The F.S., one of the most modern rail systems in Europe — thanks to the massive electrification and modernization program given such impetus under Benito Mussolini ("Il Duce got the trains running on time") — had instituted a policy of steam elimination that is almost unique in the annals of railways. General practice calls for gathering the last active steam locomotives at several division points, allowing expensive servicing facilities — such as water plugs, coal elevators, and boiler shops — to be shut down in the areas where steam is extinct, and relieving from duty or transferring boiler-makers, firemen, and other steam specialists. Not in Italy! There were few "steam districts," although the bulk of surviving steam was concentrated in the north. Rather, two or three, or a dozen locomotives were to be found at most big engine houses, from the Alps right down to Sicily, and most diesel and some electrified lines still saw at least a percentage of tonnage and passengers, as well as switching, handled by steam, well into the 1970's.

Shuffling through the Po Valley, on the light rails of the secondary line between Codogna and Cremona, an affable little Mogul of pleasing proportions leaves behind the setting winter sun and the village of Reggione.

All photos this chapter, Ron Ziel (except as noted)   43

The somewhat bland, yet trim 2-8-0's were as ubiquitous as the Moguls and were found on both local passenger and freight services. On the line from Rome, northwestward to Viterbo, just two round trips daily were made in steam — one passenger and one freight. At dawn on cloudy November 15, 1972, Consolidation No. 740-123 (*upper left*) brings her ancient coaches through small farms south of Viterbo. A few hours later, No. 740-009 (*left*) pauses with a freight at Cesano di Roma. Incredibly unorthodox would be just about the most polite description that could be given to the Franco-Crosti boilered 2-8-0's of class 743. With two supplementary smokeboxes alongside the boiler, leading to twin smokestacks ahead of the cab, these machines were quite sophisticated and efficient — but their appearance was irreconciliable to railway enthusiasts. One of the engines, No. 743-398, waits beneath mainline catenary with a freight train at Codogna (*above*) while 2-6-0 No. 625-058 barrels by with a local *passeggero* bound for Cremona. The date was January 5, 1971, and Italy and France were still digging out of the worst blizzard in many years. Equally grotesque sister No. 743-339 (*right*) is serviced at the terminal at Brescia while awaiting her next road call.

Howard Serig

Much of the F.S. and the industrial switching was performed by tank shunters. On the nationalized railway system, many of the large port facilities, such as Genoa, Mestre, and Savona, are switched by private contracting firms, which own a variety of little locomotives for this purpose. At Mestre, the port and rail center of Venice, the *Raccordi Ferroviari Mestre* lined up trains for F.S. Moguls and Consolidations, using their own stable of 0-6-0T and

0-8-0T iron ponies. On January 4, 1971, 0-8-0T's No. 10 (*above*) and No. 7 work the interchange, while R.F.M.'s big 0-6-0T, No. 2 (*below*) shunts nearby industrial yards. A type already ancient when many were taken over by the advancing Allies in World War II, the 835 class 0-6-0T was still a standard shunter thirty years later. No. 835-191 (*right and above*) pauses by a baroque town house in Cremona and later puffs through the yards.

Austria

Fairly erupting in a spate of condensing vapor, No. 93.1311 (*left*) one of approximately one hundred remaining Austrian 2-8-2T's, blasts away from a tiny country station near Gross Haslau after taking aboard a solitary passenger bound for the end of the branchline at Martinsberg. This was New Year's Day in 1971 and the Arctic atmosphere lent credence to the local sobriquet for this windblown plain: "Little Siberia." Situated on a narrow pass at the base of the overwhelming Alpine summits that completely surround it, the triangular-shaped railway yard at Hieflau harbored veterans of World War II, such as 2-10-0 No. 52.6966 (*above*) and 2-8-2T No. 86.476 (*below*) left behind by the Germans twenty-six years previously.

Left, Ron Ziel   *Two photos, Mike Eagleson*

What the locomotives of Austria lacked in aesthetic appeal, they made up for in fascinating operations in some of the most spectacular mountain scenery in Europe. By 1973 the surviving classes of steam totaled about 350 engines. Approximately half were the omnipresent German 2-10-0 *Kriegslokomotiven* of World War II; the rest were native Austrian types. Whether climbing the Alpine passes, crossing the central plain, or gathered about the dank sooty sheds of Vienna, the steam power of the *Österreicher Bundesbahnen* never lacked for quaint variety, right up to the end of the steam era. Unfortunately, however, the Ö.B.B. locomotives were, along with those of Greece, perhaps the filthiest in all of Europe.

Above, Mike Eagleson

Native Austrian tank engines were characterized by water cisterns which were mounted so that their tops were half-way up the boilers and ran most or all of the way from cab to smokebox front. Of decidedly Germanic appearance, most had been converted in later years to flat Giesl ejector smokestacks. The three most common classes were of the 2-8-2T, 4-6-2T, and 4-6-4T wheel arrangements. On May 8, 1971, 4-6-2T No. 77.09 (*above*) pulls her train of six diminutive four-wheel cars through Freistadt, bound for the Czechoslovakian border. Shown here are engine terminal scenes in the winter of 1970–71. No. 93.1413 (*left*) is in her stall in the Hieflau roundhouse. Wien Ost and Wien Nord depots, right in Vienna, were well supplied with steam power, even toward the end. At Wien Ost, a 2-8-2T, a 2-10-0, and a 4-6-2T (*upper right*) await calls to duty outside the old running shed. Dark, dirty, forbidding even by day, the shed at Wien Nord (*right*) housed all three major tank engine classes beneath its steam-saturated rafters. While typical Austrian practice favored double smokebox doors, such as those on the 77 class (*above*), the newest Austrian-designed steam locomotive, class 78 of 1931 (*lower right*), showed German influence in its entire front-end design. Here, No. 78.613, one of the last to run, rolls from Wien Nord shed. By this time, the entire Ö.B.B. steam roster consisted of tank engines, except for the German wartime 2-10-0's, which were the standard freight power.

Four photos, Ron Ziel

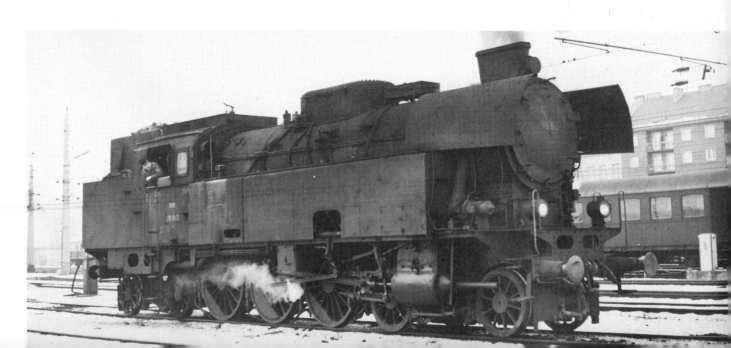

A few of the narrow gauge lines which once operated in what is now northern Austria and southern Czechoslovakia remained in service, with some steam working, well into the diesel era. At Gmünd, on the border of Czech Bohemia, narrow gauge 0-8-0's with four-wheel articulated tenders such as No. 399.04 (*left*), were still coaled by hand in the 1970's as they worked the narrow gauge lines into the surrounding countryside. The adjoining standard gauge yard was shunted by 2-8-2T's and an immaculately maintained Czech 2-10-0 crossed the border to interchange freight several times a day, making this one of the most interesting of locomotive centers in Austria. Crossing the Zwettl River trestle, 2-8-2T No. 93.1392 (*above*) recalled a childhood fantasy of primeval railroading as the minikin puffer and her lilliputian brace of four-wheel coaches traversed the winter storyland of arcadian Austria. Even in the final time of steam, rarities such as the 1907 six-coupled shunter (*right*) occasionally blundered out onto the mainline. Owned by an industrial firm at Judenburg, she paused for water at Selzthal on December 29, 1970, on her way to Eisenerz for a general overhaul.

Three photos, Ron Ziel

Undoubtedly the most fascinating and increasingly renowned railway operation in the Alpine reaches is the Erzberg Railway, a cogwheel rack line which runs from Eisenerz to Vordenberg in central Austria. Threatened since the early 1960's with "final solutions" ranging from dieselization to abandonment, this venerable line seemed destined to continue in operation with its fleet of 1890-era 0-6-2T's. These locomotives have four cylinders, each pair controlled by a separate throttle. In normal operation, the outside cylinders power the driving wheels on rails in the usual manner — known as adhesion. Whenever a locomotive reaches a section of track with the rack installed between the running rails, the second pair of cylinders is cut in to operate the gear that engages the rack, adding enormous power as the little engines tackle gradients of 7 percent. The principal freight is iron ore (*upper left*), though

Three photos, Ron Ziel

54

Above, Mike Eagleson

regular passenger service was still maintained in the early 1970's. During the winter, this delightfully archaic operation carried thousands of young skiers (*lower right*) to the slopes at Präbichl. In addition to the fourteen 0-6-2T's, three enormous 0-12-0T's, Nos. 197.301-303, designed in 1912, were used as pushers and yard switchers. No. 197.302 (*lower left*) switches the yard at Eisenerz on December 30, 1970. The pair of German-designed 2-12-2T's, built in 1941

and the largest and newest of the three rack classes, were retired about two decades later, but one has been preserved. 0-6-2T No. 97.209 (*above*) works a two-car passenger train in May 1971, hardly showing her seventy-eight years of grueling torture on the rack. Unlike the far-ranging adhesion engines of Ö.B.B., these matchless rack lokies may survive for generations to come if those sensitive souls who are battling to preserve the line succeed in their calling.

West Germany

Thoroughly devastated in World War II, the German railway system was subsequently stripped of much of its remaining serviceable motive power by the occupying nations. This situation gave a reprieve to many older classes of locomotives, which were rebuilt, and necessitated an extensive postwar program of new steam locomotive construction. Germany's phenomenal economic recovery also extended the steam season, since rapidly rising demands for new motive power — diesel and electric — absorbed these units as soon as they were built. Even in the final steam years, laws limiting long-haul trucking extended the service

Ron Ziel

Mike Eagleson

lives of many *Deutsche Bundesbahn* steam locomotives. Although steam elimination was scheduled for the mid-1970's, a thorough and well-planned preservation program ensured the *Eisenbahnfreunde* (railfans) at least two examples of every major class of locomotive for preservation, many in operating condition.

Contrasts on the D.B. On December 28, 1970, when only ten of an original total of nearly five hundred 078 class (ex-Prussian State Railways T-18) 4-6-4T tank engines, dating back as far as 1912, were still active, No. 078-474 (*left*) pulls a local south from Horb. Shuffling along near Dettingen, with a consist of six-wheel coaches, the old Prussian veteran belies her age as she traverses the edge of the legendary Black Forest. The sunlight glistening on the Mosel River at the cathedral-anointed city of Koblenz (*above*) also silhouettes on 023 class 2-6-2 — Germany's final steam locomotive endeavor — as she departs with a passenger train for Trier on September 20, 1969.

57

Undoubtedly the locomotives most sought after by foreign photographers were the 01 Pacifics and the 038 Ten wheelers, the latter being the oldest class to survive into the final decade of German steam. Built in the thousands for the Prussian State Railways as class P-8 between 1906 and 1924, only a score lasted into the '70's in West Germany. By mid-1972, they were all retired in both German states. Of the hundreds sent to other European countries, built by license in those countries, or left behind by the German Army in World War II, an appreciable number survived the DB 038's only in Rumania and Poland. In the winter of 1970–'71, 038-722 (*left*) brings a five-car local into Hechingen, and 038-553 (*lower left*) switches a dead 2-10-0 beneath the catenary at Tübingen. Like the venerable 078's, these patriarchal 4-6-0's ran their last miles south of Stuttgart in the state of Württemberg. Their throaty exhausts, for six decades muffled by the grand pine trees of the Black Forest, have fallen forever silent. In the twelve years between 1928 and 1940, fifteen different manufacturers erected 520 chunky little 064 class 2-6-2T light passenger engines. Prolific throughout Germany, these engines saw service on branch line locals and switching terminals. By 1970, a few dozen were still operating in central Germany. Reposing in the roundhouse at Aschaffenburg, No. 064-106 (*right*) waits out a winter afternoon before taking a passenger train down the scenic Miltenberg branch. At Tübingen, 064-518 (*below*) shares the house tracks with 038 class 4-6-0's and 050 class 2-10-0's. The little tank engines were to outlive the classic Ten wheelers.

Four photos, Ron Ziel

Steam with variation is characteristic of the *Deutsche Bundesbahn*, with identical locomotives of the same class being rebuilt, reboilered, and some converted to oil-firing. Such a class was the 042 Mikado. Heavy 2-8-2 No. 042-356 (*left*) straddles the inspection pit at Rheine in the damp early evening of December 20, 1970. At the same terminal, smoke from the stack of sister engine 042-096 (*right*) shows that the hostler is raking her fire. These locomotives, built between 1936 and 1941 to burn coal, have since been rebuilt with welded boilers and converted to oil fuel. The 42 class designation had originally been assigned to 2-10-0's built in the World War II era.

Two photos, Ron Ziel

Mike Eagleson

Mike Eagleson

Mike Eagleson

The last class of D.B. steam locomotive to emerge from the German erecting halls was the 023 class 2-6-2, 105 of which were built from 1950 to 1959. The 065 class, although designed later, was completed several years before the final 023 order. As medium-sized passenger power, these compact, trim Prairie-types handled mainline and secondary trains. The prototype of the series, No. 023-001 (*above*) wheels her consist past an abandoned signal tower near Königshofen on October 3, 1972. No. 023-069, traversing the since-electrified Mosel River line at Lehmen (*left*) on September 10, 1969 passes a Gothic castle on her way to Trier. At Crailsheim, the hostler boards an 023 (*right*) to take the engine into her stall in the roundhouse, while the fireman in the cab completes the timeless ritual of dumping the fire into the ashpit.

Mike Eagleson

Ron Ziel

The swift high-stepping prima donnas of the fastest German express trains were the several classes of Pacifics which rode on high driving wheels with slender spokes. As entire classes of modern steam power reached retirement, these big 4-6-2's were among the earlier ones to go, a result of the important expresses being dieselized. By 1970, just four classes of Pacifics, totaling about fifty engines, were still in service. The 01 Pacifics, such as 01-062 (*opposite, bottom*) leaving Meriz in 1969, were generally considered to be the epitome of the successful classes of heavy passenger power. A lighter version of the 01 class, 003's, such as the one riding the turntable at Ulm on the day after Christmas in 1970 (*opposite, above*), still saw service on the lines radiating from that south German city. Only fourteen coal-burning 011 class 4-6-2's, a 1939 three-cylinder development of the 01, were still in steam when the 011-091 (*opposite, center*) awaited a call to action at Rheine. Fat-stacked 012's, the oil-burning conversions of earlier 011's, held down high-speed passenger runs on the westernmost north-south line. 012-059 brings her train from Emden southbound through Meppen in 1969 (*above*), and 012-063 fairly erupts oil smoke and steam (*below*) as she thunders by the Reckenfeld tower on the Rheine-Münster main the following year.

Mike Eagleson

Ron Ziel

Undoubtedly, the last active engine on the D.B. will be a 2-10-0 of 044 or 050 class. These were the famous *Kriegslokomotiven* — war locomotives — of which an incredible total of more than 10,000 were built between 1939 and 1945! Thousands of these were operated throughout the vast areas of the Third Reich, and even in the final years of European steam, they were the mainstays of the rosters of eight countries and could be found in several more. Traversing the Mosel Valley, Decapod No. 044-117 (*above*) crosses the high river bridge, about to enter Prinzenkopf Tunnel at Bullay. A sister 2-10-0 (*below*) wheels iron ore through the Mosel vineyards to the industries of the Saar. A 2-6-2, No. 023-023, rides the Crailsheim turntable (*upper right*) virtually surrounded by 50 class 2-10-0's. With *Fröhe Weihnachten* (Merry Christmas) lettered on the smoke deflector of the lead locomotive, a brace of 52 class (same as the 50's) war engines (*right*) work the grade at Neuenstein on December 23, 1970. *Following spread:* Fresh from Lingen Shops, a big 2-10-0 assists another on an ore train at Meppen, in the early autumn of 1969. Two thirds of German steam in the last years were the *Kriegslokomotiven* — they will be the last and the best remembered.

Ron Ziel

Above and following spread, Mike Eagleson

Mike Eagleson

68

Three photos, Ron Ziel

The last D.B. design were the eighteen 2-8-4T's of the 065 class, built between 1951 and 1954. The first of the series (*above*) was also the last steam engine assigned to Darmstadt, in December 1970, when two D.B. snow shovelers watched her board the turntable. The last 065 built (*upper left*) was also the last to run, shown here heading down the Miltenberg branch from Aschaffenburg less than two years later. Old No. 078-410 (*lower left*) trundles a local near Deibingen on May 16, 1971. One of eight 0-8-0T's (*below, left*) which shunted the huge Eschweiler Bergwerks Verein complexes at Alsdorf, steams past the D.B. interchange. An affable old 0-6-0T, probably an ex-D.B. 89 class (*below, right*) — 1,345 of which were built in the quarter-century ending in 1906 — pulls an enthusiasts' special at Ottenhofen on September 14, 1969.

Mike Eagleson

East Germany

Mike Eagleson

With the partition of Germany into two separate states by the victorious Allies after World War II, the railway system of East Germany (*Deutsche Demokratische Republik* or D.D.R.) retained the original name of *Deutsche Reichsbahn* (D.R.). Although new designs in steam locomotives as well as existing classes often retained the same designations, the D.R. engines acquired a unique appearance generally more pleasing than those of the West German *Deutsche Bundesbahn*. Right through the 1960's, the D.R. followed up the new designs of the 1950's with an extensive rebuilding program of older classes. Despite this ultramodern steam fleet, the planned date of elimination of steam on both German systems was about the same — sometime in 1977.

Of all the handsome classes of D.R. steam, the most celebrated were the 01 Pacifics, some of which had running board skirts as well as the smooth front ends and skyline casings common to all. Fortunately, they were assigned to Berlin expresses and operated directly into West Germany, as well as throughout the D.D.R. No. 530 (*above*) climbs out of Bebra, West Germany, bound for Berlin on September 12, 1969, while at the other end, No. 517 races through the Grunewald in West Berlin (*left*) on a foggy, gray December 13, 1970.

All photos, this chapter, Ron Ziel (except as noted)

Two of the elegant D.R. 01 Pacifics — Germany's finest in steam — rest in the evening at the D.B.'s Bebra round-house between runs. It was between this terminal and the East German border, just eight miles away, that so many Western photographers, unhindered and unquestioned, managed to capture the goings of these engines. The elegant 500-series 01 4-6-2's were not entirely new, having been rebuilt from older standard 01's like those which survived on the D.B. into the 1970's. They received newer, larger boilers, and some even got boxpok driving wheels, among other improvements, so they were virtually new locomotives. An original, unrebuilt D.R. 01 (*below*) races a passenger train down the well-maintained mainline between Berlin and Dresden at better than 60 m. p. h. on October 5, 1972.

Mike Eagleson

Like the *Bundesbahn*, the *Reichsbahn* wound up with a vast number of World War II engines, in addition to older classes of 2-10-0's. The D.B. did little to change the appearance of these 52 class machines, except to replace the rounded tenders with higher-capacity square ones in most instances. The D.R., on the other hand, retained the original tenders on most of the 52's, while rebuilding some of the engines. Except for the addition of the standard D.B.–D.R. smoke deflectors, adopted at the end of the war, the locomotive pulling a heavy mineral freight near Wainsdorf (*left*) had changed little since the adoption of the standard design in 1942. A sister, No. 52 5199 (*lower left*), had been updated with a Giesl Ejector, creating a flattened stack. This postwar development greatly improved the efficiency but did nothing for the aesthetics of the locomotives to which it was applied. This particular specimen is pulling an ore train out of the yard at Haldensleben, northwest of Magdeburg. A very successful three-cylinder rebuilding of a number of old Prussian G-12 2-10-0's at Zwickau in the early 1960's resulted in the 58 class. On October 7, 1972, the twenty-third anniversary of the establishment of the D.D.R., No. 58 3028 (*right*), flying a red flag and a D.D.R. flag on her smokebox, pulls a mixed freight out of Wünschendorf. As the sun set in the hazy industrial sky of the southwestern section of the D.D.R. near Leipzig on that same day, oil-burning 2-10-0 No. 44 0223 (*below*) rolls westward near Etzelbach.

Perhaps the most distinctive feature of the postwar classes and rebuilds of D.R. steampower was the flat-topped Heinl feedwater heater, which was set into the smokebox, ahead of the stack, accentuating the massiveness of the front end while improving the generally formidable appearance of these typically Teutonic locomotives. These Heinl front ends appear on all three classes of standard gauge power on this spread. Larger and heavier than the D.B. class of the same number and wheel arrangement, the D.R. 65's were introduced in 1954. No. 65 1052 was one of several of her class (*above*) which switched the yard at Triptis. Eighty-eight of this class were built, as well as twenty-seven of the smaller version of the 2-8-4T, designated class 83. The newest classes appeared in 1956 and were built until production of new steam power ceased in 1960. One of these, the 23 class (later redesignated 35 class), was proliferated to 114 engines. Pulling a medium-weight passenger train, one of these 2-6-2's, No. 35 1055 (*upper right*) shows her power at Wainsdorf. The other new class was, appropriately, a

2-10-0, designated a sub-class of the 50 types, of which ninety-two were built. Another recent rebuild of an older class was the 41 class 2-8-2, one of which was pulling a freight through Tangerhütte (*right, center*) on October 8, 1972. Although the political decision was made to end development of new steam classes by 1960 (as in the United States and Russia previously), the East Germans wisely upgraded hundreds of older engines to supplement the mere 322 new standard gauge engines built since the war. A few narrow gauge lines survived around Dresden, but were being phased out at an alarming rate in the mid-1970's. Operating from the Dresden suburb of Radebeul Ost to Radeburg, small, modern (Karl Marx Works, formerly Orenstein & Koppel, 1953) 2-10-2T's were the final development of German narrow gauge steam traction. In the autumn of 1972 one 2-10-2T, 99 1783, hauls standard gauge freight cars on narrow gauge flats (*right*) down a city street and another, 99 1784, pulls a passenger train in woodland terrain (*far right*) of pure Germany, near Friedewald.

Howard Serig

Finland

Second only to Portugal in maintaining its steam locomotives in a commendable state of repair and appearance, the Finnish State Railways handily lived up to the spotless reputation of the brave little nation it so ably serves. Because steam workings were scarce during the summer tourist season and the operating locomotives relatively thinly scattered throughout the country during the winter operating times, when the severe weather made the pursuit of trains a difficult business, relatively few foreigners ventured forth to photograph the steam power of the State Railway, *Valtion Rautatiet.* In the 1970's, as steam became increasingly scarce in western Europe, however, attention belatedly turned to Finland, where several classes in six wheel arrangements continued to work through the forests and along the lakes of this extra-ordinary land. The motive power itself was impressive and nearly a perfect blend of German and American design in appearance. The most handsome features of both countries were incorporated — the clean, straight, functional lines and smoke deflectors of Germany and, on road power, the high big headlights, number boards, and strap steel pilots in the finest traditions of the vanquished steam engines of North America. In later years just fifty or sixty locomotives were under steam in widely scattered areas during the summer. With the freezing of Finland's northern seaports in November, however, scores of stored engines were fired up and used until the spring thaw to handle the increased traffic of the vital timber industry. Indications were that this delightful situation would prevail into the late 1970's, with Finnish steam outlasting that of the neighboring Soviet Union which, using the same five-foot gauge as the V.R., interchanged thousands of cars annually.

Perhaps the most elegant of V.R. steam locomotives were the twenty-two Pacifics of class Hrl, built by the Finnish firms of Tampella and Lokomo, in Tampere, between 1937 and 1957. The last two built, with roller bearings on all axles and rodding, Nos. 1020 and 1021, were also the last two in regular service. They worked daily mail and passenger trains between Pieksämäki and Kouvola until May 22, 1971, when the summer timetable bumped the duo off those runs for good. In the last winter of regular service for the big 4-6-2's, No. 1021 left her snug round-house stall at Kouvola (*below*) while an equally impressive 2-8-2, No. 1035 of class Trl, stood by. Although late afternoon by the clock, it was already dark in the sub-Arctic deepest winter as the last steam-powered passenger train in the V.R. timetable (*left*) was about to depart the ultramodern station at Kouvola. For a half century, the same engineer had operated and maintained a little Krauss 0-6-0T (*right*) built in 1909. In 1971 the line was dieselized — because the engineer retired, according to the official explanation! The 5½-kilometer narrow gauge line carries logs between two lakes at Honkataipale.

Reino Kalliomäki

Ron Ziel

During and after World War II, the Vulcan Iron Works of Wilkes-Barre, Pennsylvania, built a variety of interesting locomotives in varying sizes and wheel arrangements for the U.S. Army and for foreign governments, including an order of 0-6-0 tank engines for the five-foot gauge V.R. These 1400's, class Vr5, were later rebuilt into 0-6-2T's and the side tanks were removed. Five months after No. 1413 (*right*) was photographed shunting at Lahti in November 1970, she had been towed to Riihimäki for scrapping.

Painted a deep blue-green, the Vr1 0-6-0T's (*right*) and Vr3 0-10-0T's (*above*) work out their last miles in the yards at Kouvola in shining splendor — rare indeed for humble switching "goats." One of the exciting sights on cold (24° Fahrenheit) clear November 5, 1970, was to observe the Finnish engineers racing these shunters from one end of the yard to the other at perhaps fifty miles per hour, as if furiously seeking out more Russian boxcars to kick around!

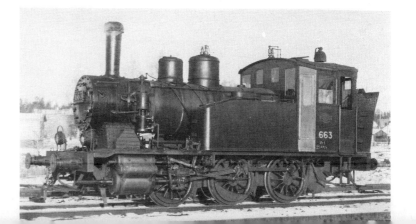

Five photos, Ron Ziel

The big, handsome Tr1 Mikados were the heaviest
class of freight engine on the Finnish Railways.
Three different builders turned out sixty-seven examples (Nos. 1030-1096) of these engines from
1936 to 1957 and winter after winter, as other
types were being hauled to the scrapyards, virtually the entire class of 2-8-2's was steamed up
to work the snowed-in countryside. No. 1086, a
Lokomo product of 1955 (*above*) pedals down to
the roundhouse at Kouvola after bringing in a
freight. Chuffing past Vr1 0-6-0T No. 662 (*below*),
a 2-8-2 heads out to couple onto a train that in-
cludes many cars interchanged with the railways
of Finland's big neighbor to the east, the Soviet
Union.

Perhaps the locomotives that will be longest remembered as being most typically Finnish are the Tk3 and Tvl 2-8-0's, many of which were modern woodburners, such as No. 1150 (*left*), built by Frichs in 1949, switching the yard at Oulu in the winter of 1971–72. These mixed-traffic Consolidations saw service in all categories, except the heaviest freight and express passenger trains. Really kicking up her heels and giving the brakeman a ride to remember, little Vrl No. 662 (*right*) speeds through a snow flurry in Kouvola, her cylinder cocks wide open, as she makes up a mainline freight. Photographer Reino Kalliomäki, a young railway enthusiast who accompanied the author in pursuit of steam on the V.R., was elected president of the Friends of the Locomotives Society, a very active organization whose dedication to preserving Finnish steam has resulted in the saving of part of a narrow gauge line. The museum train, pulled by a 1917 Tampella 2-8-2T (*lower right*) is shown crossing the trestle at Jokioinen in June 1971. Some long-term preservation of broad gauge steam is assured on the V.R., thanks to the strategic planning of the Finnish Army. In a move that should have been taken by many other countries in the interests of national defense — most notably England and the United States — the Army has requested V.R. to hold 250 steam locomotives for storage against future emergencies. Although many are coal-fired, all could burn Finland's abundant right-of-way fuel — wood — if necessary. Locomotives such as No. 1140 (*below*) are stripped of moving parts, coated with a preservative called *Dinitrol*, and sent to various Army storage depots. Eventually, twenty-eight 0-6-0T's, sixteen 0-6-2T's (not the Vulcans), fourteen 4-6-0's, ninety-eight 2-8-0's and the entire classes of Vr3 0-10-0T's, Hr1 4-6-2's, and Tr1 2-8-2's will receive this treatment. If the railways of Western Europe and North America are ever cut off from their overseas supplies of diesel oil, they can look back at Finland's wise decision to store the best of its steam power and wonder at the shortsightedness of their own strategic planners!

Three photos, Reino Kalliomäki

Above, Ron Ziel

Bulgaria

Like most lines of Eastern Europe, the Bulgarian State Railways (B.D.Z.) maintained their steam power in immaculate condition. A grimy or rusty locomotive being a rarity, indeed. In their livery of green and black, with red frames and running gear and brass appliances highly polished, they were beautiful to behold, if at times remote to photograph. Most of the B.D.Z. types were possessed of a Teutonic look, due to the aggressive salesmanship of German builders between the World Wars and the subsequent alliance between the two nations. No. 02.02, a Mikado (*upper left*), flushes out a flock of birds as she crosses the Thracian Plain near Plovdiv. A type of locomotive unique to Bulgaria was the massive 2-12-4T, built in two classes weighing 149 tons and 155 tons respectively. They were so gargantuan that their fuel and water capacities were as great as the largest British tender engines, and they

sported even larger boilers! On October 20, 1966, less than five years before the unfortunate premature withdrawal of the 2-12-4T's, No. 46.12 (*left*) works a freight north of Sofia. An express 4-6-2, No. 05.01, pulls a fast train out of Plovdiv (*lower left*) in 1969. 2-10-0 No. 14.18 (*upper right*) starts a southbound freight out of Stora Zagora the same year. An ancient 0-8-0, No. 26.32 (*lower right*), still shunted at Krupnik on August 9, 1969. The 76-centimeter narrow gauge lines featured such impressive little 2-10-2T's as No. 611.76 (the last two digits designating the gauge) being watered on August 12, 1969 by her fireman at Kneza (*right*) while passengers board the cars. Replacement of steam was proceeding rapidly and by the 1970's, many of the best classes were gone and the predominating survivors were the omnipresent German *Kriegsloks*.

All photos, Frank Stenvall

For many years, the steam locomotives of the *Polski Koleje Panstwowe* (Polish State Railways) were a complete mystery to all but those who actually rode the trains of the P.K.P. As late as 1970, British enthusiasts caught in railway yards with cameras received jail sentences, so the thousands of engines still active in this big central European country were a tantalizing prize — to be admired from afar. Then, with the government of Edward Gierek, followed by the visit of U.S. President Richard Nixon to Poland in 1972, rapid improvements began taking place. When this apprehensive author-photographer arrived in Warsaw on October 10, 1972, he was amazed beyond all expectations to find awaiting him a letter of permission to photograph anywhere on the P.K.P. — apparently the first such blanket permit ever granted!

Polish railway officials then assisted him in ferreting out the rare classes and even traveled with him, tearing down the few remaining barriers. Hostility had turned to the traditional warm hospitality of which the Polish people are justifiably renowned. The results are shown here — the first batch of good action photos of modern, postwar P.K.P. power to be published outside of its magnificent homeland. The significance of these gleaming, superbly maintained locomotives should not be underestimated, for steam is scheduled to remain in service into the 1980's. If the current official attitudes prevail, Poland will become the last stronghold of steam in Europe, attracting photographers by the hundreds, their quarry adding considerably to the enchantments this ancient land has always held for fascinated foreigners.

Poland

All photos this chapter, Ron Ziel, October, 1972

The standard tank engine of Poland is an enormous general purpose 2-8-2T of class Tkt-48, designed and built to a total of 194 in the late 1940's and '50's. Although classified as freight power, these adaptable machines are to be found on Warsaw suburban runs, the appointed rounds for No. Tkt-48 194 (*upper left*) at Modlin, and local passenger services, a task for Tkt-48 175 (*left*) tackling the grade out of Nowy Sacz, bound for Tarnow in the mountains of southern Poland. It is on the latter lines that P.K.P. locomotives are expected to wage the final battle of steam in Europe, so scenes such as Tkt-48 95 doubleheading a German wartime 2-10-0, No. Ty-2 285 (*above*), on a mixed freight out of Nowy Sacz, may become familiar to many railway photographers. The "T" designation is for *Towarowy*, meaning "express." The "k" designates tank engine in the form of *Kusy*, which means "abbreviated, chopped off or truncated" in Polish — a charming and colorful abstract for these tenderless machines!

Like so many countries of eastern Europe ravaged by World War II, Poland wound up with a large roster of German 52 class Decapods — more than 1,900 (including some later purchased from the Soviet Union), a goodly number of which seem to have survived into the mid-1970's. Designated Ty-2, Nos. 159 and 674 (*right, above*) had no sooner halted at Grybow when sister 881 arrived with another freight. No. 702 (*right, middle*) battles upgrade at Stary Sacz, and Nos. 124 and 1077 (*lower right*) prepare to move a night freight out of Warsaw Gdansk station in the capital. While many Ty-2's retained the *Reichsbahn* smoke deflectors, some had the distinctive standard P.K.P. high-mounted wing type. In the 1950's, 232 big, brutish 2-10-0's of class Ty-51 were added to the roster as the final development in Polish steam design. Ty-51 186 (*below*) brings a heavy freight into Modlin, as the fireman of Ty-51 57 (*bottom*), armed with oil and rags, wipes down his big Decapod with typical P.K.P. regard for appearance, and the engineer catches a nap on his cab seat.

The P.K.P. operated a number of older (but no less well maintained) classes, wartime foreign engines and modern types. The handsome 2-8-2 (*upper left*), No. 5 of 180 examples of the Pt-47 class, is shown with a local at Wileñska station in Warsaw, a few months after diesels had bumped her off Bialystok expresses. Five hundred American Army World War II 2-8-0's went to Poland in the late '40's and hundreds survived as the standard switch engine, class Tr-203. No. 368 (*center left*) pauses beneath new catenary at Pomiechówek as Tkt-48 11 brings in a commuter local. Designed to replace the aging Prussian P-8's, 2-6-2 No. 01-49 39 (*lower left*) works north from Debicia. The 116 01-49's were successful, but the venerable Teutonic P-8 Ten wheelers of class 0k-1 (*right and above*) at Ostroleka look healthy, indeed, in the autumn of 1972. An early example of the light 2-10-0, designed at the close of World War II, No. Ty-45 24 (*below*), doubleheading into Ostroleka, was one of 448 built to help spur the devastated economy back to life.

89

If all of the standard gauge activity remaining in steam on the P.K.P. boggled the minds of railway connoisseurs, the extensive narrow gauge lines were an antiquitarian's delight. In western Poland, many of these lines were German before 1945, including military railways. The lines originally built in what is now eastern Poland were privately owned, but like the German portions, were nationalized into the P.K.P. in 1945. A few unmarked specimens still operate on spurs into industrial areas and probably are under the control of those particular factories. Most of the narrow gauge lines are of 750 mm. width, with some 600 mm. and 785 mm. surviving. In addition to prewar Polish and German locomotives, Russian stragglers and even a few oddities from the Baltic States, probably brought down by the *Wehrmacht*, have been seen. 120 Px-48 750 mm. gauge 0-8-0's were built in the first half of the 1950's, and some of them are the regular power on the incredible little commuter line which operates right on the shoulder of the main road from just outside of Warsaw Wileňska depot up to Radzymin. At sunset, No. Px-48 1752 (*above*) switches around her train at Wileňska. One hundred and fifty miles southward, 2-8-0's, such as No. Pxa-9 1811, Polish-built by Chrzanow in 1931, worked at such bucolic back-country villages as Kolosy (*opposite page*) on an extensive network of lines that, regrettably, suffered an influx of shiny new diesels. Connecting with this line at Kazimierza Wlk., a tiny 0-6-0T, bearing no P.K.P. markings (*below*) but numbered 1521, shunts freight between a factory complex and the P.K.P. yards. What undiscovered treasures in steam still remained to be sought out, even at so late a date, in the proud nation of Poland!

Soviet Union

It is most regrettable for railway steam locomotive historians that the sudden easing of tensions between the Union of Soviet Socialist Republics and the Western nations — particularly the United States — came so late in the era of steam in Russia. Even then, when the possibility of photographing Soviet steam was seized upon by a few American and British enthusiasts, the formidable bureaucracy of Russia greatly compounded matters. Indeed, a photographer who spent two weeks stalking the Soviet Railways in the early 1970's was fortunate to get pictures of a dozen live engines and to get his film out of the country. In addition to the obvious dangers inherent in photographing the strategic railways of Russia, the system was nearly dieselized and electrified by 1970, with just 6 percent of the trains hauled by steam, according to official sources. Thousands of former road engines had been downgraded to yard switching chores which made reaching them even more difficult, and written permits were nonexistent. It is perhaps understandable that Soviet authorities looked askance at the innocence of railway photography. In any event, those who ran the risk of being charged with espionage were not disappointed by what they saw, for the always rare and oft beautiful steam locomotives of the vast land that reaches from central Europe to the Pacific Ocean were breathtaking to the fortunate few who got to them.

The Soviet railways possessed classes of locomotives that totaled well up into the thousands. Constructed over a period of decades, some operated for more than a half century. The most celebrated class, however, was the magnificent green, red, and yellow P36 4-8-4, which only came along in 1950. These were destined to pull the trains traveled by the foreign tourists who began descending upon the Soviet Union in the 1960's. The first one — a passenger design from the outset — was erected in 1950, and because of the policy of subordinating passenger power to freight, it was not duplicated for four years. Approximately 250 were erected by 1956, when the decision to phase out steam was made, cutting short a program that undoubtedly would have resulted in thousands of these great engines being built. Indeed, the L-class 2-10-0 freight locomotives, built in the thousands, were their contemporaries, and the last year of steam construction saw five hundred engines built — most of them heavy 2-10-2's, which may have already been extinct by the 1970's, none having been reported by foreign observers. The highest known numbered P 36 was No. 0250, generally believed to be not only the newest Soviet 4-8-4, but the last mainline engine built for domestic service as well. Cleaner and more proudly maintained than her sisters, and with an unusually friendly crew that was not camera-shy, P36-0250 was assigned to Trains 1 and 2, *Russia* — the trans-Siberian express. On November 16, 1970, No. 0250 took over Train No. 1 from sister 0126 (*above*) at Skovorodino, and the following day P36-0041 paused at Chita (*left*) in central Siberia.

All photos this chapter, Ron Ziel (except as noted)   93

With the electrification of the Moscow – Leningrad mainline, the 4-8-4's were relegated to the Ukraine, where the last ones in European Russia operated from Lvov in 1971. Ever-dwindling numbers continued to do passenger chores in Siberia, such as the two (*left*) and a sister (*lower left*) taking water at one of the turret-like water towers of the early 1900's, this one being at Zubariovo. The standard medium passenger engine, built in the thousands for forty years ending in 1951, was the high-boilered Su class 2-6-2. By October of 1972, Su 215-61 (*right*) was the last steam engine to switch around the ancient fortress-city of Novgorod. Sisters 251-00 and 252-34 (*below*) backed through the coach yard at Riga, Latvia, to pick up their train on a bitter cold New Year's Day in 1971.

Below, Howard Serig

American-built locomotives for export overseas were shipped in the tens of thousands during the first half of the Twentieth Century. The largest single order ever consigned to one foreign nation, however, were the approximately 1,900 Ye class 2-10-0's sent to the Soviet Union under the lend-lease program during World War II. Quite similar in appearance to the famed "Russian Decapods" of World War I, many of which remained in America and were converted to standard gauge after the October 1917 Communist Revolution suspended deliveries after 875 had been shipped. Most, if not all, of the later series spent their service lives in Siberia. In the last years of Soviet steam, none were reported seen west of the Urals, but hundreds still worked as the standard switching class and on local freights on the Trans-Siberian route's easternmost segment, between Irkutsk and Khabarovsk. By 1971, the line-ups of these Ye class war engines in some of the larger depots along the line totaled fifty to one hundred locomotives, with perhaps a thousand of them retired in the past few years and awaiting dispatch to scrapyards. The Ye's were credited with having averted a grave locomotive shortage when the Russians could produce few of their own and lost thousands during the German occupation. Like many of her sisters, No. 4160, a June 1945 Baldwin graduate (left), finished her days as an industrial shunter. Although many Soviet railwaymen were camera-shy, the engineer of 4160 (below), on November 12, 1970, seems quite content in his comfortable all-weather cab as he switches an asphalt plant in Irkutsk. Ye 3068 was photographed from Train No. 1 four days later (lower left) as she basked in the winter sun at Taldan, just a few miles from the Chinese border. With the possible exception of the World War II U.S. Army 2-8-0's, the Russian Decapods constituted the largest class of identical steam locomotives ever built in the United States.

"I am for the steam locomotive. I am against those who imagine that we shall have no steamers. It is a robust machine, stubborn, and will not give up. However, one should not idealize the steam engine. I should say straight out that although in the future the steam locomotive will remain, the leading locomotives will be electric and diesel."

— L. Kaganovich, Minister of Ways and Communications of the U.S.S.R. at a planning conference in 1954, *when he was urging the construction of 6,000 new steam locomotives during the 1956–60 five-year plan. His ideas were rejected and he was later purged.*

The standard tank switching engine, built in large numbers, was the P class 0-6-0T, one of which is shown (*above and lower left*) in the obscurity of the factory yards of Irkutsk in November 1970. The most numerous class of steam locomotive ever built, surpassing even the German war engines, was the Soviet Railways' E class 0-10-0. More than 11,000 were erected in a 30-year period, ending about World War II. No. 788-75 of this type (*lower right*) was one of the last active steamers in the Ukraine, when photographed at Novograd Volyns on October 22, 1972. By the end of World War II, the Soviet Railways had taken into stock about 1,750 German 2-10-0's, many already converted to five-foot gauge by the *Wehrmacht*. A few survived until the last days as class Te, including No. 4777 (*upper right*) with the University of Riga in the background in December 1970, and No. 3266 (*middle and lower right*) at Narva two years later. As oilburners, all of the locomotives in northern European Russia had huge cylindrical fuel tanks added to the original tenders.

Above photo, Howard Serig

The finest — and last — in Soviet steam strutted their best well into the 1970's, although the original replacement program, laid down in 1955, had called for the total phase-out of steam by 1970. The huge grain purchases from the United States, which began deliveries early in 1973, also may have prolonged the life of the magnificent 4-8-4's and 2-10-0's for yet another season, for virtually all of those millions of tons of feed grains and wheat would have to be shipped from Russian ports to the vast interior by rail, probably creating heavy motive power and car shortages. Although the Soviet government has not released the production figures, as many as 5,200 of the standard postwar class L Decapods may have been built during the late 1940's and early '50's. As the finale of steam drew nigh, they were the most numerous of the surviving classes in western Russia. L-0773 rolls a heavy freight (*below*) through the electrified yards in Minsk in the winter of 1970–71 and L-1817 brings a solid block of oil cars (*left*) southbound from Chudovo on the secondary line between Leningrad and Moscow at twilight on October 17, 1972. The massive P36 4-8-4's were influenced in their design by the L class 2-10-0's, although the obvious inspiration of the Southern Pacific Railway's famed "Daylight" GS class locomotives is also evident in such striking examples as P36-0250 (*upper right*) as she waited to move Train No. 1 westbound from Skovorodino, Siberia, on November 16, 1970. Decapod No. L-0395, resplendent in green, red and black and with the Cyrillic "USSR" emblazoned on her smokebox at Narva (*right*) in the winter of 1972–73, was indicative of the pride and the individualistic decorative sense of Soviet railway workers right up until the final demise of steam.

Below, Howard Serig

Czechoslovakia

Czechoslovakia, carved out of the Austro-Hungarian Empire after World War I, had a great variety of odd classes of locomotives. Many were obsolete at the time. Much modernization had progressed before the Nazi takeover in 1938 and by the end of World War II, German and Hungarian power had found its way onto the rails of the *Ceskoslovenske Statni Drahy* (Č.S.D.). Although a few examples of the older types survived into the 1970's, the preponderance of modern Czech power was exemplified by two classes of superb 4-8-2's — regarded by many *aficionados* as the handsomest locomotives in Europe — plus the most sophisticated and well-proportioned tank engines in the world; the 477 class 4-8-4T's. In freight, the ubiquitous ex-D.R. 52 class Decapods of class 555 and postwar Skoda-built 2-10-0's of class 556 still shared the rails with older and lighter 534's and the homely 2-8-0's of 434 designation, the survivors of which were relegated to yard switching in later years. Prewar 464 class 4-8-4T's also worked some branchline passenger services. Č.S.D. motive power was renowned for mechanical excellence and the cleanliness standards which prevailed were noted throughout the world. All of this was to pass before the end of the 1970's.

The 4-8-2's of class 475 were built by Skoda between 1947 and 1951. After completing 147 for the Č.S.D., twenty-five more in the continuous number series were erected for the railways of North Korea. During a spring thunderstorm, one of the trim, high-boilered 475's No. 160 (*left*), lopes along a fill, crossing the fields at Lulec. Rounding a curve near Libstat on May 24, 1971, a typically well-maintained Mountain, No. 475.197 (*above*), wheels a passenger train in the grand manner. The year 1948 saw the introduction of the heavier 498 class. One of the first, No. 498.007 (*right*) roared through a country station north of Tabor with a fast train from Prague. The final express class, completed in 1955, were the heaviest 4-8-2's, such as 498.111 (*below*) taking water at Leopoldov. Prior to the liberalization of domestic policies under Alexander Dubcek, all Č.S.D. locomotives sported big red stars on their smokebox fronts. During the "Prague Spring" all of them were removed, and as late as author Ziel's first trip to the country in December 1970, not one engine was seen wearing a red star. Just five months later, when author Eagleson's party arrived, all of the red stars had been reinstated. The last vestige of the old order had returned.

Three photos, Mike Eagleson   Above, Ron Ziel

103

A 475 class Mountain, No. 1125, emerges from a tunnel near Adamov (*upper left*) in May 1971. For rapid acceleration and frequent stopping needed in suburban and local passenger service, the Č.S.D. began development of an extremely successful series of 4-8-4T classes, beginning with the two-cylinder 464.0 type in 1933. One of the earlier examples, No. 464.016 (*above*), built in 1936, was assigned to local service on the branchline between Pilsen and Klatovy, where she still shows fine form thirty-six years later as she rounds a curve through a cut at Kokšin. The most modern and certainly the handsomest of tank locomotives operating during the twilight of world steam, were the three-cylinder 477.0 class, built 1950–55. A splendid specimen sported a Communist slogan and Czech and Soviet flags (*below and left*) as she pulled a heavy passenger train near Stare Mesto in the Spring of 1971.

Three photos, Mike Eagleson   Above, Ron Ziel

The standard Č.S.D. heavy freight engine in the last quarter century of steam was the Skoda 556.0 Decapod, 510 of which were built from 1952 to 1957. Among the handsomest of European freight locomotives, the 556.0's not only worked over the Č.S.D., but ran into neighboring Austria and Hungary as well. Traversing the now electrified mainline up the Danube Valley from Budapest toward home rails, Č.S.D. 2-10-0's (*left and opposite center*) show their heels on the Hungarian State Railways in the Winter of 1970–71. Power for the grade! No. 556.071 (*below*) works upgrade near Stare Mesto, and 556.0183 (*upper right*) acts out the classical role of the hard-working pusher engine as she emerges from a tunnel on the steeply ascending spiral curve at Svermovo on May 21, 1971. Thirty cars ahead, a sister 556.0 handled the chores on the point. The standard light 2-10-0's were of the 534 classes, built between 1923 and 1947 in many varieties, culminating in the 534.03 series after World War II. A 1945 Decapod, No. 534.0329 (*lower right*), trailing a six-wheel mini-tender normally reserved for older, lighter sisters, left Klatovy on October 27, 1972.

Mike Eagleson

Two photos, Ron Ziel

Above, Mike Eagleson

More than four hundred *Deutsche Reichsbahn* 52 class war engines were taken into Č.S.D. stock as the 555 classes. Some survived until the last of steam; No. 555.3278 (*above*) still worked near Pestinok on May 19, 1971. The 434 classes of 2-8-0's were as ugly as the 4-8-2's were beautiful. With backset cylinders, outside connecting pipe between domes, chopped three-axle tenders and generally stubby appearance, the oft-renumbered 434.245 (*below*) sat out her last active years in the engine terminal at Benešov, a half century after her 1921 debut. A 464 class 4-8-4T (*upper right*) chuffs through the foggy fields near Červeñe Poříči in 1972. A more modern 477 class of the same wheel arrangement (*right*) passes the locks on the Vltava River in Prague as she brings a suburban passenger local into the capital. Silhouetting a semaphore in her exhaust (*far right*), a 534 class 2-10-0 leaves Tabor. A gleaming Mountain type, No. 475.194 (*bottom right*) traipses through the springtime fields near Stitna Popov in central Czechoslovakia.

Hungary

Hungary was another country that wound up on the losing side in both World Wars, being stripped of locomotives (in some instances, entire classes) as well as real estate. Although the *Magyar Állam-vasutak* (Hungarian State Railways) once ran compound mallets, streamlined tank engines, and a pair of experimental 4-6-4's, the mainstay of the M.A.V. roster in the last years of steam were the three classically Hungarian and two foreign classes shown in this chapter. Appreciating successful designs, M.A.V. motive-power men were not adverse to dusting off old blueprints and ordering new locomotives nearly identical to ones constructed thirty years previously. Interestingly, all passenger service was handled by native engines while the two classes of American and German war locomotives are the principal freight power. Justifiably the most renowned of Magyar locomotives, the magnificent 424 class 4-8-0's, were dual-service machines. Found all over the country, they hauled anything a yardmaster felt like coupling behind them. So respected are the 424's that when a team of Hungarian Secret Police were questioning the author after he had been arrested for taking locomotive pictures, the interrogator informed him, with an unmistakable tone of pride, that the 424's had won major international awards of excellence after their 1924 debut. Then the serious questioning was resumed!

American Military Railway Service veterans of World War II would be amazed to know that hundreds of the bald-faced wartime standard U.S. Army 2-8-0's survived into the 1970's in Eastern Europe. After the G.I.'s were through with them in France, it was not deemed advisable to return them to America, since they were hardly suited for civilian railways that were still ordering some steam — virtually all being monstrous super-power engines — but already were committed to diesels. With the Army having no further use for them, 510 of the Consolidations wound up in Hungary, where, almost thirty years after the war, they were still the principal freight mover, with hundreds puttering about the country displaying their U.S. Army Transportation Corps and builders' plates as well as huge red stars on their smokebox doors. On a cold morning in December 1970, No. 411.441 (left) is silhouetted against the cloudless sky as she works a train of ore cars near Nagykáta. Almost two years later, on October 28, 1972, No. 411.283 (above) switches boxcars and 411.446 (below) works a freight beneath the catenary in electrified territory at Tatabánya. The engines' appearance had changed little in three decades, except for the fitting of typically Hungarian smokestacks and repositioning of the air pumps to the sides of the smokeboxes. Although electrification was advancing steadily, the M.A.V. had not extensively dieselized and, being equipped with a hefty roster of postwar steam power, apparently had elected to keep these locomotives in service through the 1970's.

All photos this chapter, Ron Ziel (except as noted)

If there is a class of locomotives that may be said to be the embodiment of typical Hungarian design, it is the 324 class of 2-6-2 general purpose locomotive, used extensively on mainline local passenger trains, as well as branchline freight and passenger service. An original batch of 355 were built from 1909 to 1913, followed by 143 more in the first two years of World War I. Another 397 followed by 1923. Twenty years elapsed before rebuilding resumed, with nearly 100 more added in the 1500 block of numbers, making a total of close to 1,000. Possessed of virtually a comical appearance, with enormous "ashcan" stacks, plain front ends, short eccentric rods that give the engine a "limping" look when in motion, and stubby six-wheel tenders, the 324's changed little through their sixty-five years of service. Another idiosyncracy of the class is that no two seem to have identical skylines, with steam dome, sand box, air reservoirs, feedwater purifier, and other appurtenances being moved about and encased in many different manners. The Brotan-boilered 324.934 (*lower left*), passing the impressive water tower at Kisber with a branchline passenger train, was built in 1916. She appears no older than 324.1581 (*left*), probably thirty years her junior, shown here struggling upgrade with a capacity freight at Acsa, on one of the many secondary lines that were almost the exclusive preserve of these homely Prairie engines. Two years previously, in 1970, No. 324.1546, another World War II-era example (*below*) brings a local through Zebegeny, north of Budapest, on its way to the Czech border. This scenic line through the valley of the Danube, which even saw big Č.S.D. 2-10-0's on freights run right down to the Hungarian capital at one time, has since been electrified, ruining one of the most photographic steam lines in Europe.

There are a number of border-crossing locomotives between East and West in Europe, such as ex-German 52 class M.A.V. 2-10-0 No. 520.030 (*left*) about to return to Hungary from Wiener Neustadt, in Austria, in January 1972. In October of the same year, a light 2-6-2T, No. 375.1009 (*below*) brings a train of four-wheel mail and passenger cars through Nyul on her northbound hop to the main line depot at Györ. Diminutive as these engines were, an even lighter version, class 376, was still active on some branchlines. If the 324 and 375 classes cast doubt on the ability of the M.A.V. to produce aesthetically pleasing locomotives, the majestic 424 class 4-8-0 should settle the issue. Perhaps the most beautiful example of this wheel arrangement ever built, the 424's were as mechanically perfect, powerful and reliable as they were handsome. Not all of the estimated 365 of these engines remained in Hungary; some going to Czechoslovakia, Yugoslavia, Russia, and even China. Three immaculate examples of the finest in Magyar steam are shown in December 1970 at Farmos (*lower right*) and Gyömrö (*upper right*) on the central plains and at Zebegeny (*above*) pulling a heavy freight virtually on the banks of the Danube River.

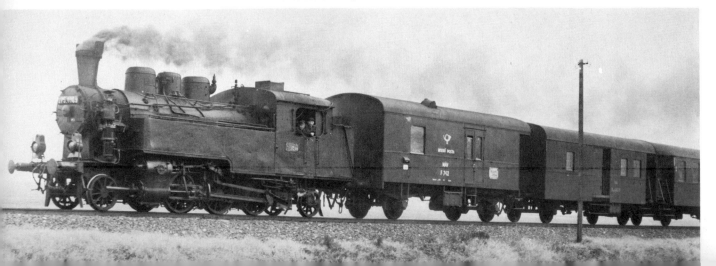

Yugoslavia

The motive power situation of the Yugoslav State Railways is immensely complex. Between the formation of Yugoslavia after World War I and the turmoil of World War II, the *Yugoslovenske Pzavni Železnice* (J.Ž.) inherited an incredible hodge-podge of standard and narrow gauge motive power from Austria, Hungary, the old Balkan states of Bosnia, Herzegovina, and Serbia, as well as German locomotives of both conflicts and American machines after 1945. There were no "typical" traits to unify the appearance of J.Ž. steam engines, but the widespread class of ex-Serbian 2-6-2's built in Germany were distinctively Yugoslav. Smoke deflectors were an on-again-off-again proposition, with locomotives of the same series boasting specimens with and without them. By the 1970's, with steam still doing much work on the standard gauge, many narrow gauge lines were being closed, equipped with diesels or converted to broad gauge, but much steam activity survived.

As the author was being led away by two policemen after being captured while photographing locomotives at the yard limit in Lapovo, he managed to get a picture of a Serbian 2-6-2, No. 01-092 (*left*), as she sped by with a little local down the Morava River Valley on November 29, 1970. Duly impressed with the Hungarian 424-class 4-8-0, the J.Ž. had acquired fifty-two of them when Tito's feud with Stalin cut off further deliveries. A final order of ten engines was sent in 1955, however, including No. 11.058 (*right*) arriving with a freight at Ruma. The feature that made these engines so impressive was the large smoke deflectors. The older sister running near Tinja (*below*), bound for Tuzla, is rather awkward in comparison, bereft as she is of the big "elephant ears." A brace of 05 Pacifics (*bottom*) bring the southbound *Marmara Express* into Sićevačka Gorge, in southern Yugoslavia on August 14, 1969.

Three photos, Ron Ziel   Below, Frank Stenvall

Shielded against the freezing winter winds by a wrapping of hemp, a water column frames a German Decapod (*left*) at Lapovo. Along with Greece, the only place to find World War II U.S. Army 0-6-0T's in regular service in the 1970's was in Yugoslavia, where 105 went, as class 62, after the war. They were well received and an additional twenty-five of modified design were built in the late 1950's. No. 62–644 switches a coal mine loader (*lower left*) a few miles south of Tuzla, far removed from her G.I. days. A handsome Mikado works the line between Ljubljana and the Italian border (*below*) with an express in the winter of 1971. Meanwhile, back in Lapovo, the author managed to wriggle through a circle of about twenty railway shopmen and police who were rummaging through his passport to photograph some of the locomotives at the engine shed. He got about a half-dozen color and black-and-white pictures before the Yugoslavs realized what had happened, including a shot of the fireman checking the valve gear of 2-6-2 No. 01-101 (*right*) while ex-D.R. 2-10-0 No. 33.126 prepares to move a freight southward. With No. 97-020 on front and No. 97-014 pushing, a passenger train on the 760 mm. gauge line at Donji Vakuf (*lower right*) tackles the 4.5 percent grades.

Two photos, Ron Ziel

Below, Howard Serig

259
6

Ron Ziel

Below, Frank Stenvall

Rumania

The concord between Germany and Rumania strongly influenced the design of the steam power of the *Caile Ferrate Romane*. The Rumanians frequently chose proven Teutonic designs — most notably the ever-popular Prussian P-8 4-6-0, the G-8 0-8-0, the G-10 0-10-0, and the *Deutsche Reichsbahn* 50-class 2-10-0 — then built them in large numbers under license. The most famous of Rumanian locomotives, however, were seventy-nine imposing 2-8-4's that were duplicates of the Austrian 214 class of the 1930's, only thirteen of which were constructed. Other German and Austrian locomotives, as well as Hungarian, American, and Russian power was adopted by the C.F.R. as Rumanian borders experienced massive contortions after both World Wars, greatly affecting the locomotive population on several occasions. The Rumanians did design their own power as well, frequently turning out very handsome and massive locomotives. By the mid-1970's steam was rapidly disappearing as diesel and electric units arrived in large numbers.

A big Decapod, having gained momentum out of Sebes, deposits a heavy trail of vapor (*upper left*) on the bleak, damp evening of November 25, 1970, as she crosses the Walachian farmlands. The Magyar legacy, in the form of locomotives such as ex-Hungarian Railways 2-6-2 No. 324.567 (*left*) bringing a freight into Timisoara, was evident until the last years of C.F.R. steam. Nearly one-fourth of Rumanian steam power consisted of the Prussian-designed 0-10-0's, more than 800 of which were ordered between 1921 and 1942. A glistening example, resplendent in shining brass trim (*above and below*), switches at Lotru.

All photos this chapter, Ron Ziel (except as noted)

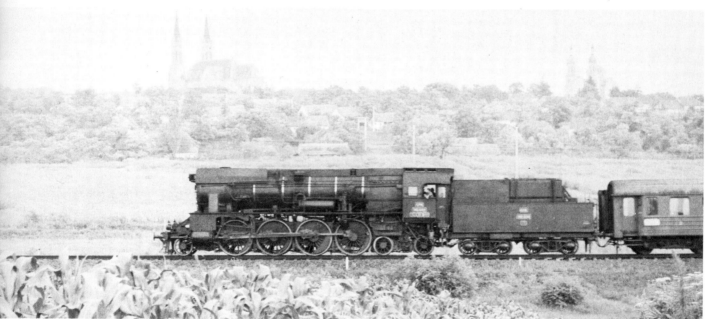

Photo above, Frank Stenvall

The first Prussian G-8's to arrive in Rumania were built in 1914. Seven years later, the C.F.R. ordered seventy-six more during a two-year program. The first engine of this latter batch, No. 40.037 (*upper left*), switches the big passenger yard at Sibiu in November 1970, still peppy after nearly a half-century of hard service. One of the 2-8-4's designed in Austria, No. 142.072 (*left*), wheels a passenger accommodation past the cathedrals of Vinga, between Timisoara and Arad, on July 19, 1970. A postwar 2-10-0 based on the D.R. 50 class, No. 150.091, scoots under leaden skies (*lower far left*) bound for Deva. Another Hungarian holdover was 2-6-2T No. 375.929 (*lower near left*), the yard "goat" at Lotru. A larger 2-6-2T of 1939 design, class 131, was the newest C.F.R. type. The author was literally hiding in the hedgerow as he snapped the magnifi-

cent P-8 Ten wheeler (*above*) leaving Pitesti, while the depot policeman's attention was momentarily diverted by the steamy antics of the departing local. Mounting a gold star to complement the golden bands around her stack, 0-10-0 No. 50.200 wheels a freight out of the yard at Pitesti (*below*), while sister 50.560 makes up another train. One unique feature of C.F.R. power was the practice of burning a mixture of coal and oil, hence the oil tank on top of the tender water tank, directly behind the coal bunker. The many railway *aficionados* around the world who collect locomotive number, class and construction plaques would go berserk with their clandestine tool boxes in Rumania! The average C.F.R. steam engine and tender combination carry no fewer than eighteen red and white embossed cast plates, as these photographs clearly show.

Greece

Although many steam engines were serviceable and kept in use, the vast majority of train movements were diesel-hauled by the 1970's. For example, ex-U.S. Army 2-8-0's Nos. 525 and 392 (*left*) are hot at the Drama shed on November 1, 1972, but not one steam train moved on the mainline to Thessaloniki that day. More active is No. 530 (*below and right*) as she struggles to move heavy tonnage out of Ghefyra the following day. No sooner had the exposure of No. 535 and her crew in the Thessaloniki yard (*lower right*) been accomplished, when she moved her freight out into the night, toward Athens.

Steam was dying a very lingering death in Greece, and even in the early 1970's reports that the show was just about over proved happily untrue. Granted, in the previous decade entire classes of extremely old and, in some instances, very new locomotives had disappeared, leaving just a few stragglers among the many classes of mostly Austrian-influenced steam power. In Thessaloniki alone, more than fifty engines had been left derelict for so many years that it was doubtful that either the track or the locomotives could take the strain of movement to a scrap yard. The survivors were mainly of three classes: the big Italian-built 2-10-2's of the mid-1950's; the 0-6-0T U.S. Army shunters (many of which had been converted to 0-6-0's by the simple expedient of removing the panion tanks and adding a tender from a retired engine); and the standard U.S. Army S-160 class World War II 2-8-0's. Long ago, it seems, the Hellenic State Railways cared about the appearance of its motive power, but in the last years, Greek steam locomotives had acquired one common trait: virtually all of them were sooty, grimy and streaked with rust.

All photos this chapter, Ron Ziel, November, 1972.

The biggest and most impressive locomotives on the Hellenic State Railways were also the heaviest standard gauge steamers in Europe, as well as one of the most trouble-plagued classes ever built. When the Greeks needed a heavy freight locomotive in the early 1950's, they turned, for some inexplicable reason, to Italy, which had not turned out a mainline steam locomotive in more than twenty years and had never constructed a heavy-hauler of such monstrous proportions. The result was predictable; the firebox was woefully inadequate and the big 2-10-2's were always short on steam. Their weight, already at the limit, precluded the necessary modifications that would have brought the engines up to full potential, although they were withdrawn immediately after delivery and some improvements were made. The dire predictions that they would not survive their first decade were wrong, probably because those responsible for the debacle could not bring themselves to admit to such a blunder and withdraw the locomotives. Greek enginemen had to put up with them right up to the end of steam. While the design of the twenty Santa Fes was inadequate, there was no questioning their formidable demeanor. No. 1006 (*left*), stating the case in her family's defense, brings a long freight through Diavata on the last lap of a run from Athens to Thessaloniki. Twenty U.S. Army 0-6-0T's, numbered 51–70, replaced an older class of 2-6-0T's as the standard switcher. No. 52 (*lower left*), working the yard at Drama, retained her side tanks a quarter century later. Others, such as No. 58 (*below*), the regular shunter within the Thessaloniki limits, had their tanks removed and tenders added to reduce their weight on the light rail of marshaling yards. Survivors of the twenty-seven G.I. 2-8-0's delivered after the War and twenty-five more bought third-hand from the Italian State Railways in 1959–60, will probably close out the saga of steam in Greece by the mid-1970's.

Denmark

Steam was banished from most mainline duties quite early in Scandinavia, and Denmark was no exception, although seasonal needs and test runs of stored locomotives put off the "final elimination" of steam, in a technical sense, for many years. By 1965, only seventy-two steam locomotives were still active, and all regular service on the *Danske Statsbaner* petered out about 1970. One of the last active locomotives, D.S.B. S class 2-6-4T No. 737 (*left*), blasts a Roskilde-bound passenger accommodation through the Copenhagen suburbs on September 7, 1967. On March 12th of that same year, Mogul No. 19 (*above*) pulls a farewell excursion over the Lollands-banen shortline in southern Denmark shortly before it closed down. The biggest and best remembered of D.S.B. express engines were the E-class Pacifics. Ten were purchased from Sweden after the 1936 mainline electrification retired them, and the Danes built twenty-five more themselves in the mid-1940's. No. 974, an original Swedish 4-6-2 (*upper right*), wheels freights after being bumped from varnish runs. After the death of King Frederick IX, an event occurred that impressed not only his mourning countrymen but railway enthusiasts everywhere. The king had always loved steam locomotives, and he had requested that steam power be used to carry him to his final resting place. D.S.B. officials worked hard to ready two stored Pacifics, Nos. 978 and 994, and on a cloudy January 24, 1972, the Royal Funeral Train (*lower right*) made its melancholy journey. One result was a great resurgent interest in Danish steam. Since the funeral, the Pacifics have been kept busy in excursion service. On March 3, 1972, 4-6-0 No. 963 (*right*) is shown near Hedehusne. The following month the resplendent and very rare four-cylinder 4-4-2, No. 917 (*below*), is at Aarhus station to head up an enthusiasts' special.

108

Nina Carlsson

Torben Nielsen

Jens Koefoed

Norway

Steam workings wound down slowly in Norway at about the same rate as in Denmark, leaving the *Norges Statsbaner* with about one hundred operable locomotives by 1966, most of which were gone a few years later. Norwegian power had to be good steamers to work the long grades. Since all coal was imported, they had to be economical as well, and double use of steam through compounding was the rule. A typical class that met the stringent requirements was the 30c 4-6-0, such as No. 469 (*above*) in the Gudbrandsdalen Valley near Hamar in July, 1968. Perhaps the archetypal N.S.B. steam locomotive was the slanted-cylinder 4-8-0, such as No. 434 (*below*) and No. 229 (*upper right*), both in the vicinity of Hamar. The best of Norway's steam power was the big 2-8-4 class which included No. 463 at the Stören depot. Called *Dovregubben* ("Dovre Giant"), the class worked the Dovre Line between Oslo and Trondheim. When the Swedish Railway Club toured the N.S.B. on May 10, 1970, one could still stage a photo run-past featuring steam trains of two gauges passing the cameras simultaneously. This scene (*lower right*) is at Höland.

Two photos, W. E. Dancker-Jensen

Sweden

A complete lack of domestic coal and oil reserves, coupled with abundant hydroelectric power, wrote an early end to Scandinavian steam, and by the mid-1960's more than half of the *Statens Jarnvagar* (Swedish State Railways) had already been electrified. From then on, simple economics dictated the use of railcars on branchline passenger runs and diesel power for freight work. Still, some extra moves, such as a

trainload of Swedish Army tanks (*below*), still rolled behind steam — in this case, inside-cylinder E-class 0-8-0 No. 1042 — as late as 1967 on the Malmö–Revinge line. The F-class Pacifics were among the finest locomotives to run on the S.J., but were sold to the Danes after the mainline electrification program of the 1930's. In 1963, the Swedes repurchased one of them, restored her to the original livery and number, and began running excursions with her. On May 29, 1966, No. 1200 (*right*) puts on quite a show at Furuvik, beneath the

Frank Stenvall

same catenary which had run her off the property — and out of her homeland — thirty years earlier! More involved in military matters than its pious pronouncements would indicate, the Swedish government even kept the number of steam locomotives in *storage* a military secret. For many years during the winter, gleaming examples of S.J. power were fired up and run on locomotive test trains. On February 20, 1968, 4-6-0 No. 1008 and 0-8-0 No. 1538 (*above*) doublehead out of Jönköping on such a move. Two years later, No. 1926 (*lower right*), one of the twenty beautiful little S-1 2-6-4T tank engines built in 1952 as the newest of S.J. steam, runs her annual test between Bollnäs and Orsa.

Three photos, Erik Sjoholm

Rhodesia

The locomotive design of the Rhodesian Railways, like that of most other former British colonies, was heavily influenced (and usually erected) by the mother country. Pilots, large headlights and ample cabs were concessions to American practice. As in South Africa, the Beyer-Garratt, a highly successful though radical departure from conventional design, found favor in Rhodesia; indeed, it was an ideal machine in much of Africa, Australia and in parts of Europe and South America, but was never tried in North America.

A 16A class 2-8-2 + 2-8-2 Garratt, built by Beyer Peacock in 1953, departs with the daily Shamua mixed train (*left*) from the capital city of Salisbury in October 1966. In June 1971, a 15th class 4-6-4 + 4-6-4 (*lower left*) heads southbound out of Wankie with copper ore. Built by North British in 1912, 4-8-0 No. 86 (*above*) spent her later years as the shop switcher at Bulawayo. Steaming out of Mateking, South Africa, for the long run across Botswana to Bulawayo, 19th class 4-8-2 No. 336 (*below*) pulls a train of Rhodesian blacks who work the Johannesburg mines.

All photos this chapter, Victor Hand

Traversing the railway that is neighboring Zambia's lifeline to the Indian Ocean ports, one of the most modern of steam locomotives, a 1957 20th class 4-8-2 + 2-8-4 No. 727 (*above*), loaded with coal and machinery for the copper fields, passes Sambawizi Station en route to the Zambia Railways interchange at Victoria Falls. A 12th class 4-8-2 (*left*) pulls a coal drag south from Wankie in 1971 during the waning years of Rhodesian steam activity.

Turkey

The private railways of the old Ottoman Empire were amalgamated into the Turkish State Railways (*Turkiye Cumhuriyeti Devlet Demiryollari Isletmesi*) between 1927 and 1936, bringing with them a varied background of motive power of British, French, and — most numerous — German ancestry. The new T.C.D.D. continued to favor German design in the form of standard Decapods built in 1937, with post-war deliveries of German *Kriegslok* 2-10-0's, as well as similar designs erected in Britain and Czechoslovakia in the late '40's. To complete the esteem held by the Turks for this wheel arrangement, they ordered eighty-eight enormous semi-streamlined Decapods from the United States in 1947 and in 1955 finally purchased forty-eight secondhand German 44-class 2-10-0 locomotives from the French.

All photos this chapter, Ron Ziel, November, 1972.

With more than 600 active steam locomotives surviving into the mid-1970's, Turkey was one of the most desirable of the world's steam countries to study. Approximately one-third of the locomotives came from Germany, with similar amounts from Britain and the United States, as well as a few stragglers from France and elsewhere. The suburban services out of Izmir, northward about fifteen miles to Ciğli, were in the charge of the class of six 2-8-2's (46.101-106) built by Robert Stephenson in 1929. Although only three engines operated this shuttle service at a time, they appeared to be of a far greater number as they ran back and forth, half the time tender-first (*above*) due to the absence of turning facilities at Ciğli. The first of the series, No. 46.101 (*upper left*), her face bashed in and sooty, lays down a veritable smoke screen as she rolls her commuter train along the shore of Izmir Bay. At Irmak, a gathering place of big power in central Turkey (*upper right*), an American 2-10-0, a German-designed and Czech-built engine of the same wheel arrangement, and a pilot-equipped German 0-10-0, No. 55.007, work in the junction yard.

139

The most celebrated of all train-watching places in Turkey was the level crossing of the two mainlines right in the middle of Izmir. As late as the winter of 1972–73, no fewer than twenty-two separate steam movements, involving locomotives of six different wheel arrangements, could be seen crossing this diamond during the 7–8:30 a.m. rush hour on any weekday. One of the Stephenson 2-8-2's (*above*) is on her way from Bashmahane, the main Izmir terminal. A few minutes before, one of the last of the 1912 Humboldt Consolidations, No. 45.127 (*below*), clattered over the crossings on the southbound line from Alsancak station. The passenger terminal shunter at Basmahane, just

a half-mile from the crossing, was a delightful little Stephenson 0-6-2T (*upper right*) whose builder's number, 3420, nearly coincided with her road number: 3412. One of the oldest T.C.D.D. engines and also one of the last to remain in service in the capital city of Ankara was 0-6-0T No. 3328 (*right*) whose builder's plate reads: *Maschinenfabrik Esslingen, Emil Kessler No. 2909, 1897*. Nearly three decades newer were the eight 4-6-4T's of the renowned Prussian T-18 design, built by Henschel for the Turks in 1925. At least one of them, No. 3705 (*lower right*), could still be found, simmering in shining splendor and with red and yellow trim and brass in a fine polish, at Manisa.

An unusually handsome class of 2-10-2 was created during the 1930's by an expedient of using what appeared to be a conglomeration of miscellaneous spare parts lying around the erecting halls of several German builders. By combining the boilers designed for an earlier delivery of 4-8-0's, cylinders and running gear from the Prussian G-10 0-10-0 and the addition of two-wheel trucks fore and aft, twenty-seven of the big Santa Fe types, including No. 57.014 (*upper left*) leaving Izmir, performed well right into their fifth decade. Forty-three German 52-class war engines, including No. 56.531 passing Ciğli (*left*), were delivered new to the T.C.D.D. during World War II. Also German in design, but Czech-built by Skoda in 1949, 2-10-0 No. 56.125 (*above*) heads north from Irmak with a work extra. One of the standard U.S. Army locomotives of World War II was the "Middle East" 2-8-2, which saw extensive service in Greece, Iran and Turkey. In addition to the twenty-nine sent directly to the T.C.D.D. from America, the Turks purchased twenty-four more from the Iranians in the mid-1950's. No. 46.204, a March 1942 Lima engine, rests on the coaling track at Irmak (*right*) while a switch tender fills the well of an oil-burning switch lamp. In the unique numbering system of the T.C.D.D., the first numeral indicated the number of driving axles and the second the total number of axles under the locomotive, with the remaining three digits being the engine number. Thus 2-8-2 No. 223 was numbered 46.223, having four driving axles and six axles total. Tank engines used a similar four-digit system.

One of the last American-built products of the steam-lining fad of the 1930's and '40's to see regular service in the world, the eighty-eight big 2-10-0's erected by the Vulcan Iron Works of Wilkes-Barre, Pennsylvania, a firm normally associated with small industrial tank switchers, continued to turn in impressive performances on the T.C.D.D. right into the last years of steam. Based in eastern Turkey, their westernmost outpost was Irmak, where five were assigned. They were rugged and powerful and presented a handsome appearance as sort of a "vest-pocket" version of the Southern Pacific's famed *Daylight* 4-8-4's, with skyline casing, solid pilot, and curved-top welded tenders. No. 56.318 (*left*) worked up the grade on the mainline between Irmak and Ankara.

An added attraction of the Vulcan Decapods was that they usually operated two to a train out of Irmak, either doubleheading or with one on the point and the other pushing. At Lalabel (*upper left*) a doubleheader, led by No. 56.374, moves toward Ankara. Doing battle with the gradients of the Elma Mountains, two Vulcans push and pull a freight (*below*) through the way station at Kurbagli. The branchline from Irmak to the Black Sea port of Zonguldak was a preserve of the American engines. South of Kalecik, Nos. 56.387 and 56.314 (*above*) doublehead through the barren, uninhabited desert. A splendid example of Turkish engine maintenance was the American "Middle East" 2-8-2 No. 46.227 (*lower left*), waiting on the passing track at Lalabel for a pair of Vulcans to arrive.

Mozambique

While generating considerable on-line traffic, the *Caminhos de Ferro de Mozambique* is in the enviable economic position of being basically a bridge route. Indeed, the only way for the considerable rail tonnage of landlocked Malawi, Rhodesia, and — in darkest Africa — Zambia to reach Indian Ocean ports is via the rails and behind the steam power of the C.F.M. The very nature of the line, long, heavy freights operated over stiff gradients with a considerable traffic density, necessitates the use of big, modern locomotives on the trunk lines. The engine rosters of Mozambique include some of the rarist of the world's remaining steam power, such as the last known 4-4-2 Atlantic types still in regular service and a series of 4-6-4+4-6-4 Garratts. Heavy on Henschel, bullish on Baldwin, conservative on Canadian and enriched by English builders, the C.F.M. offers a grand variety of big and little locomotives for the venturesome enthusiast.

All photos this chapter, Mike Eagleson, June 1971

Atlantics running to the Indian Ocean are the finest treasures of the C.F.M. The 1923 Henschel 4-4-2 quadruplets handle all passenger trains from Nampula to the ocean-front towns of Nocala and Lumbo, on the 42 inch gauge northernmost trackage of C.F.M. Lively No. 813 poses in the Nampula yard (*below*) during a night of the full moon and later departs (*upper left*), bathed in the reflected glow of the lunar sphere, with the 3:30 a.m. train to Lumbo. Sister 811 switches Nampula (*lower left*) and brings a train (*above*) through the jungle toward the terminal.

Like most countries of southern Africa, the heaviest of motive power on C.F.M. rails are Garratts. A 921 class 4-6-4 Garratt (*below*), a veteran of both the Sudan and the Rhodesian Railways, leaves Beira while behind the shops one of five 4-8-2 Garratts (*below*) built on license by Henschel in 1956 awaits heavy repairs. These 971 class locomotives are the largest and newest of C.F.M. power and in their maroon and gray livery are the only engines not painted black. The United States is well represented in Mozambique by such power as unusually handsome Porter 2-6-2 No. 572 (*upper right*) and Baldwin 2-10-2 No. 200 (*lower right*) at Lorenco Marques. French-built 2-8-2 No. 613 (*center right*) rests at Nampula after bringing in a freight from the interior. Terrorist activities in the countryside precluded extensive lineside photography.

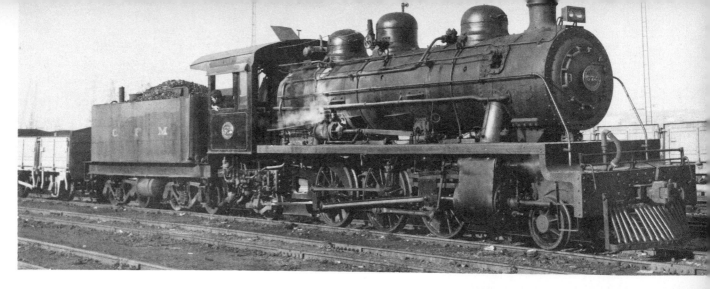

Two classes of extremely handsome Pacifics handle the passenger duties out of the main southern terminal of Lorenco Marques. Big, smoke-deflectored Henschel 4-6-2's, such as No. 332 (*above*), run into the Republic of South Africa to connect with Johannesburg expresses. Local passenger duties are in the charge of Baldwin cap-stacked Pacifics of which No. 305 (*left*) is a classic example. Also riding the Lorenco Marques turntable, post-World War II Henschel 2-10-2 No. 253 (*below*) looked formidable indeed for a 3′ 6″ gauge engine. The Lorenco Marques shunting force consists of 2-8-2T's and 0-10-0T's. 2-8-2T's Nos. 81 and 95 (*right*), the latter assisted by 0-10-0T No. 62 on the rear, are making up trains for their bigger brothers of the high iron.

Vanishing with the awesome mainline steam power they so closely supported are the coal tipples, water towers and the big-time steam shops. These repair facilities possessed common traits around the world, and the sprawling works at Lorenco Marques could just as well have serviced mainline power at Altoona, Cheyenne, Swindon, Nevers, Eveleigh or Junin. The cavernous interiors supported overhead traveling cranes with capacities of 200 tons or more. The largest locomotives were swallowed up in deep repair bays with drop pits and scores of skilled laborers sweating among myriads of spare parts. Here, too, the term "loud as a boiler factory" is vividly evident in the cacaphony of air hammers, rivet guns, machine tools, swearing boilermen and lunch whistles that echo to the grimy clerestory windows. Occupying the repair bays (*above*), a conglomeration of large and small engines undergo heavy repairs. Built in Canada in 1948, the newest of eight immense 4-8-2's (*left*) gets her valves set. A skilled machinest (*lower left*) checks the piston valve diameter of a new replacement cylinder casting. Among the running gear parts of 2-8-2 No. 614 (*below*) are the big locomotive headlights that have unfortunately been replaced by twin sealed-beam lamps of disastrous aesthetic appearance. The prototype Canadian Mountain of the C.F.M., No. 701 (*right*), is lifted off her wheels by the shop crane.

South Africa

The Republic of South Africa, where white men
and black locomotives rule supreme, is one of the
three most important countries in the bleak future of
steam. Only here and in China and India will a
semblance of mainline steam operations remain after
1980. Until quite recently the continuing rise in traffic
absorbed new diesel and electric locomotives as soon
as they were delivered, causing displaced steam to be
immediately allocated to other power-starved lines.

With the arrival of new management in 1972, the
announcement was made that the 2,400 remaining
steam locomotives of the South African Railways
would be withdrawn from service by 1986, that orders
were being placed immediately for 395 new diesels,
and that the electrification program would be ac-
celerated. For a few years at least, the S.A.R. will
continue to provide the biggest show readily available
in steam.

Climbing the steady grade through Kraankuil, a 25NC class 4-8-4 (*left*) named *Zelda* maintains her pace by the virtue of semaphore blades that read "clear track ahead." A brace of 23 class Mountains, the largest of S.A.R. 4-8-2's, No. 2553 (*above*) — the second in a series of 136 built by Schwartzkopf and Henschel in 1938-39 — and sister No. 3265 march out of Modderrivier. A pair of 15F 4-8-2's, with all 255 remaining in service as late as 1971 as the most bountiful class in Africa, doublehead a string of empty gondolas (*below*) across the South African veld toward Karee.

All photos this chapter, Mike Eagleson, June 1971

Developed by the Germans during World War II, the condensing steam locomotive proved to be ideal for operations through the arid central plains of South Africa. By recirculating exhaust steam back through an enormous tender equipped with fans and condensing machinery, as much as 85 per cent of the water may be saved, resulting in the range between water stops being extended sevenfold. The prototype condenser, the only one built by Henschel, 4-8-4 No. 3451 (*above*) is at Kimberly. Eighty-nine more were constructed by North British in 1953–54. In the manner of all condensers, the banjo-faced 25 class produces a peculiar whining rather than chugging sound when in operation. Leaving De Brug at sunset, 23 class Mountain No. 3232 (*below*) works a goods train. A German-built class 16D Pacific, designed by Baldwin, No. 872 (*upper right*) brings an evening commuter train out of Bloemfontein. One of ninety-five class 15CA and 15CB 4-8-2's built in the late 1920's, No. 2847 (*lower right*) simmers at Capital Park depot in Pretoria.

Two of the three S classes of 0-8-0's are represented by No. 366 (*left*), a 1928 Henschel, switching Germiston yards and S1 No. 374 (*below*), a heavier shunter, at Bloemfontein. Twelve of the latter, the only homemade steam power, were built in 1947 by S.A.R.'s Salt River Shops. Constructed just after the turn of the century, an 8 class 4-8-0 (*below*) serves as pilot at the Bloemfontein coal stage. Renowned as heavy-haulers that are flogged to the limits of their mechanical endurance, GO class 4-8-2 Garratts, such as No. 2584 (*right*), virtually monopolize the grueling gradients of the Steelpoort branch. Working toward her home shed at Lydenburg from Belfast, the garrulous Garratt howls her defiance to the skies as she conquers grade and tonnage at Tom Long.

Certainly the heaviest employer of Beyer-Garratt's unique design, South African Railways ordered them in fourteen classes of five wheel arrangements and two gauges. From 1924 to 1968 more than 400 Garratts were built for S.A.R. by nine different European companies — mostly German. Among the earlier Garratt series, and for many years the most numerous class of these machines in the world with sixty-five built by Hanomag, Henschel and Maffei in 1927–28, class GF 4-6-2+2-6-4's such as No. 2373 (*above*) working out of Welgegund survived in large numbers into the 1970's. Two-foot gauge 2-6-2 Garratts were built as late as 1968 — the last steampower to come to S.A.R. No. 140, a class NGG 16 (*below*) built in 1958, brings a train into Ixopo. Approaching Lydenburg on the Steelpoort branch, a GO class 4-8-2 Garratt (*right*) climbs toward the main engine depot.

A Krupp 2-6-2+2-6-2 GCA class (*above*), the lightest of 3′ 6″ gauge Garratts, sets out a cut of cars at Donnybrook. After coming into the Lydenburg terminal, a 4-8-2 Garratt (*below, left*) is serviced before her next call. GMAM Mountain Garratt No. 4107 (*below, right*) is wiped down at Krugersdorp. North British built 100 vest pocket 2-8-4's for S.A.R. in 1949–50. Berkshire No. 3645 (*upper right*) of class 24 steams over the ashpits at Bloemfontein. A trio of big, brutish GO 4-8-2 Garratts, Nos. 2583, 2576 and 2591 (*center right*) snooze in the sun between runs at Lydenburg. Until the electrification line crews arrived in 1972, the engine shed at Bloemfontein (*lower right*) housed more than 100 locomotives and dispatched more than 150 steam movements daily, making it one of the largest steam terminals the world had ever seen. *Following spread*: Reflecting the earliest rays of dawn, a 15F class Mountain assaults the hill at Karee with a long-distance express.

Kenya and Tanzania

Frank Stenvall

The central African nations of Kenya, Tanzania and Uganda are served by one railroad system — the East African Railways — which has boasted a roster of steam locomotives so modern, efficient and powerful that comparable diesels would cost more money to operate. The Garratts that are largely responsible for this steam success story are ably reinforced by several classes of tender and tank locomotives. Although E.A.R. operates 4-8-4+4-8-4 engines.

the massive 251-ton 59th class 4-8-2+2-8-4's — thirty-four of which were placed into service in 1955 and 1956 — are the heaviest and, at 83,350 pounds of tractive force, truly gigantic for their "narrow" 39⅜-inch gauge. Indeed, as the world's most powerful meter gauge engines, they are stronger than the original American standard gauge 2-10-4 design — the 600-series of the Texas & Pacific!

Two photos, Victor Hand

A modern 2-8-4, erected by North British in the mid-1950's, No. 3024 (*upper left*) pauses at Itigi, Tanzania, on the former Tanganyika Central, built by the Germans just prior to World War I. A 13th class 4-8-4T, No. 1308 (*left*) shunts the yards of Nairobi, Kenya, in October 1966. That same month, a 29th class 2-8-2, No. 2931 (*above*) brings a block

of stock cars into Marimbeti on the mainline from Mombassa to Nairobi. Throughly upgraded, including the application of Giesl ejectors, these older classes helped write the performance records that made the E.A.R. the envy of master mechanics on many diesel railways.

Victor Hand

Two photos, Frank Stenvall

The East African Railways employ a variety of Garratts as their prime tonnage-movers. A 56th class 4-8-2 + 2-8-4 (*upper left*), resplendent in maroon and black livery and with gold trim, awaits the call to action at Dar Es Salaam, Tanzania, in June 1971. A pair of double-heading 55th, class machines (*left*) pause for servicing at Mwatate, Kenya, in August 1970. Although of the same wheel arrangement as the giant 59th class, 60th class engines such as No. 6011 (*lower left*) are considerably smaller. This one is en route from Moshi, at the foot of Mount Kilimanjaro, to the Indian Ocean port of Tanga. Bringing a heavy goods train from Uganda, 4-8-4 + 4-8-4 No. 5811 (*above*) rolls into Nairobi in 1970. Four years earlier, a massive 59th class locomotive (*below*) climbs the grade from the Athi River with freight bound from Mombassa to Nairobi. Steam is king in the twilight of its reign!

Two photos, Victor Hand

Zambia

The British African colonies of Northern and Southern Rhodesia became independent states in the mid-1960's — the former by agreement with the mother country and changing its name to Zambia, and the latter by declaring itself independent and shortening its name to Rhodesia. Prior to that time, the Rhodesian Railways, with a long mainline and branches in both countries, had been administered by one authority. After independence, years of negotiation were required to divide up the assets (including the locomotive stock) between the two new nations, and a joint commission continued to hold title and, to an extent, coordinate operations. The Zambesi Sawmills Railways was the only line in the western part of the country, and two or three times weekly a night passenger accommodation ran from Mulobezi down to Livingstone and back. On the August 20, 1970 run, the train, powered by a trim 4-8-2 of probable South African Railways origin (*below*), was held at a remote siding for seven-and-a-half hours due to track repairs on the line — a common occurrence on this casual operation. The morning train from Kitwe, up near the Congo border, to Chililabombwe, powered by a brace of 2-8-2 Garratts, pauses for water (*right*) at a desolate tank. Some zealous Zambians apparently got a jump on the Rhodesians, when they lettered a 9B-class 4-8-0, No. 84 (*lower right*), with a "ZR" before the motive power issue between the Rhodesian and Zambian Railways had been settled. No. 84, the last of her class in service, was the mine shunter at Kabwe, the administrative headquarters and main shops of the Zambian Railways.

Three photos, Frank Stenvall, August, 1970.

India

Three photos, Paul S. Stephanus

The Indian sub-continent has always held a great — almost fatalistic — fascination for Westerners, and the railway historians of the 1970's were no exception. As late as 1970, it was estimated that over 55 per cent of the world's trains were powered by steam with this total rapidly diminishing. By the middle of the decade, however, more than 9,000 steam engines still handled half of the rail commerce of India. India's monumental, well-publicized problems apply to railway photography as well, as revealed in the December 1972 issue of *Continental Railway Journal*. "A warning: A reader who recently made a four-week tour of India using third-class rail travel, . . . found it extremely arduous physically and mentally exhausting, and consequently very disappointing. . . . Trains are often very crowded — so much so that it can be physically impossible to climb aboard — and huge contingents of people travel on the roof. In many rural areas foreigners are hardly ever seen and the appearance of an Englishman can attract a crowd in a matter of seconds, which even if not actively hostile will at best make photography impossible; in the cities the prevailing conditions under which the people live can be of such a desperate nature as to make railway photography appear somewhat irrelevant. Add to this the problem of security, of one's person as well as property, the extreme heat in certain seasons and the difficulties of finding good accommodation, and there is a situation which should make anyone contemplating a visit review his plans carefully." Even such rough countries as Bolivia and Jordan are rightfully considered mere "training grounds" for India. Yet, by the mid-1980's it may well be all that is left to the steam enthusiasts and no doubt will attract them like so many moths to a flame.

L. G. Marshall

Although mob scenes and conditions of squalor may hinder foreigners on the Indian State Railways, permits to visit sheds and yards are obtainable. Too, the overworked officials at these locations are usually very considerate and accommodating to photographers. The most famous of India's locomotives are the streamlined Pacifics, such as WP No. 7008 (*upper left*) at Saharanpur engine yard on June 21, 1969. Not allowing all that filthy water in a flooded servicing pit to go to waste, a girl takes a bath (*left*) at the Delhi Serai Rohilla yard. Steam versions of the New York subways at rush hour: a Manning Wardle 0-6-2T (*above*) No. 2 of the Futwah to Islampur line about to depart Futwah on November 15, 1970, and a meter gauge mainline train, pulled by a YP 4-6-2 (*below*), leaving Gurgaon station the previous year.

Approximately 1,000 4-6-2 Pacific locomotives remained in active service on the Indian Government Railways broad gauge lines in the early 1970's and nearly as many on the meter gauge. By far the most famous locomotives of any gauge in India were the streamlined WP class Pacifics built by Baldwin, Canadian and Montreal, in North America, as well as by Austrian and Polish firms, beginning in 1947. Two decades later, when the order was completed by Indian builders, they were the standard heavy express passenger engine. It is interesting to note that none were bought from England, the former "mother country." The March 1969 roster reports 755 WP's in service, by far the largest class of streamlined locomotive ever built. Photographer Stephanus rode these engines as well as photographed them at lineside. On June 20, 1969, he caught No. 7058 (*above*) rolling through Delhi past Humayun's Tomb with Train No. 80, the *Taj Express* in blue and white livery. Just west of Ghaziabad Junction a few days later, No. 7657 (*upper right*) heads for Barauni with the crack *Assam Mail* in tow. Although ample in size and somewhat sleek, the WP's rarely exceed 40 m.p.h. on account of operating conditions. Built as simply as possible, without such frills as automatic stokers, the 4-6-2's are regularly assigned two firemen. Leaving Delhi, the engineer and head fireman of No. 7008 (*left*) await the "highball" to move onto the main. En route from Delhi Junction to Saharanpur Junction, a distance of about 110 miles, No. 7008 (*right*) is fed coal by the second fireman. With only nine cars weighing 450 tons, No. 7008 was moving more than 1,000 passengers on train No. 19, the *Dehra Dunn Express*, when she made a flying meet (*far right*) with a Delhi-bound local on the well-ballasted double track.

Five photos, Paul S. Stephanus

Through the 1960's and on into the early '70's, the vast, sprawling Chittaranjan Locomotive Works continued to turn out brand-new steam power. When YG class 2-8-2 No. 3573 emerged from the erecting floor in March 1972, it was believed to be the newest reciprocating steam engine built for a major railroad for revenue service that the world would ever see. It is known that China was building steam in the 1960's, but until more is learned of that land of many mysteries, India holds the prize. One of the last of world steam, meter gauge 4-6-2 No. 2750 (*above*), a YP class whose boilers are interchangeable with the final Mikados, arrives at Delhi Serai Rohilla with Train No. 203 from Delhi. A 2-8-2 of class YG — perhaps the termination of the epic of railway steam power — passes a waterhole (*left*) near Gurgaon. Although small in number by Indian standards, just 154 built from 1924 to 1950 and only fifty-three surviving in 1970, British-built HPS class 4-6-0's, such as No. 24262 leaving Futwah (*below*), were widespread and generally regarded as one of the more handsome classes of Indian steampower.

L. G. Marshall

Five photos, Paul S. Stephanus

Working at Calcutta's Howrad station, an old British inside-connected 0-6-0 No. 34205, resplendent in candy cane boiler striping and star and cresent (*upper left*), waits for a clear signal from platform eleven to continue switching on June 7, 1969. A Mogul shuffles along yard trackage (*upper right*) while a sacred bull nonchalantly wanders across adjoining tracks. The third of the final classes turned out at Chittaranjan (production ceased before the Mikados) was the WL class of 2-6-2's. No. 5174 (*above*) is at Gurgaon in 1969. The erecting floor of the Chittaranjan Locomotive Works (*right*) is a busy place indeed in 1971, a year before all steam production ended, with the YG's still being turned out at the rate of six per month. Even after the last engine was given over to the I.S.R., production of spare parts, including boilers, was continued. Ample coal reserves and a desire to save precious foreign exchange by buying relatively few diesels and electrics overseas were the main reasons that the erecting bays of Chittaranjan were kept busy with steam power so late in its waning years.

Below, Howard Serig

Two photos, L. G. Marshall

Paul S. Stephanus

L. G. Marshall

India operates a vast network of narrow gauge railways, utilizing nearly 500 steam locomotives of great variety and style. Although diesels have made some inroads, a far greater threat emanates from highway competition, which has already caused some closures. Only the great local need for any and all transport assures a lease on life for the little grungers that chuff among the farms and villages and into the Himalayas. In the Eastern Railway region, a 2′6″ gauge line is operated between Nabadwip Ghat and Shantipur, where 2-4-0T No. 774 (*upper far left*) shunts on November 3, 1970. Almost at the other end of the country, a Western Railways 0-6-2, No. 569 (*upper left*), also of 2′6″ gauge, works at Nadiad. Operating from Delhi, a third line of the same gauge, the Shahdara-Saharahpur Light Railway, uses 2-6-2T's such as SS No. 2 (*center left*) arriving at Noli station. A meter gauge 4-4-0, No. 639 (*lower left*) is at Bhavnagar on the Western Railway. In 1971, an 0-4-0T (*right*) of the most famous Indian narrow gauge, the mountain-climbing Darjeeling Himalayan Railway, works into the foothills of the range that boasts the tallest peaks in the world. Illustrating the argument for continued steam operation in India, cheap labor is used to fill steel buckets (*below*) with plentiful local coal to be loaded into the tenders of modern WP's, which require little skilled maintenance.

Howard Serig

Paul S. Stephanus

Pakistan

Railway photography has always been difficult in Pakistan due to the remoteness of the country and the often stern attitude of the military government, as well as attacks by the Indian Air Force. Since most of Pakistan's motive power predates the partition of India and Pakistan in 1947, its locomotives are identical to classes on the Indian Railways; an additional reason why locomotives in Pakistan are rarely photographed. Two well-maintained examples of Pakistani power are 0-6-0's No. 2480 (left) and No. 2386 (below), on opposite ends of the same Sunday-only train. This push-pull operation is shown at Peshawar station and assaulting the grade on up to the legendary Khyber Pass.

Two photos, Paul S. Stephanus, 1972

Sudan

Some Sudanese steam survived into the 1970's providing interesting subjects for photography. At Khartoum, North British-built 4-8-2 No. 537 (*above*) leaves the main station while a 4-6-2 (*right*) runs backwards with an outbound freight. In 1967, a small Hunslet shunter (*below*) works the Atbara Yard.

Above and bottom, Takao Takada

Above, Paul S. Stephanus

Bangladesh

The newest nation in the world to join the ranks of operators of steam locomotives came along late in the era of steam — 1972 to be exact. As one country after another was dieselizing, Bangladesh and its steam engines became independent from Pakistan after the brief war between that nation and India in December 1971. Quite similar to the motive power of India, the locomotives of Bangladesh played a vital role, immediately after the war, in carrying food, medical supplies and clothing to the starving population. With so many pressing problems and virtually no foreign exchange, the government of the new country will probably rely on steam service for many years to come. Refugee children dance on the roofs of coaches (*left*) and a sacred cow ponders how to cross a servicing pit in the Dacca shop building (*middle left*) to retrieve her wandering calf. Natives with nothing to smile about surround Mikado No. 724 (*lower left*) and glower at the photographer. 2-8-2's (*right and below*) work to rebuild the devastated economy while Ten wheeler No. 253 (*bottom*) rolls a freight out of the yards.

All photos, Paul S. Stephanus, 1972

Jordan

The final days of steam came early to the nations of North Africa and Asia Minor, which stretch for 4,000 miles of sandy wastes and fertile coastline along the southern and eastern Mediterranean shoreline to the Persian Gulf. Although oil fuel was abundant, water was extremely scarce, and by the 1960's the lightly trafficked railways had been largely dieselized. One of the few exceptions was the Hedjaz Railway, operating southward from Damascus, Syria, to a point deep in Jordan, almost three-quarters of the way to the Gulf of Aqaba. In the late 1960's and early '70's, Syria was far too dangerous for Western railway photographers to enter, and its border with Jordan was frequently closed due to its distaste for the humanistic and civilized attitudes of King Hussein. Despite enigmatic Arab politics, however, the Hedaz Jordan Railway continued to operate uninterrupted south of Amman with a small, but very photogenic roster of thoroughly modern 4-6-2's, 2-8-2's and 2-6-2T's.

With her smoke plume remaining stationary above the curving track in the still desert air of early morning, 2-8-2 No. 52 (*left*) marches a local freight — the only class of train on the railroad — away from the Jordanian capital of Amman, in the background. The previous evening, the 1955 Jung-built Mikado and sister No. 53 (*above*) rested in the engine terminal, beneath the hills and minarets of the city. Also built in 1955, a Belgian 2-6-2T, No. 63 (*right*) switches at Om Alhiran.

All photos this chapter, Ron Ziel, November 12–13, 1972.

Although the author spent just two days on the Hedaz Jordan Railway, it was easy to get more than 100 action photos of two Mikados, two Pacifics, and a 2-6-2T. Amman taxi drivers will go anywhere in the country for just $15 for a twelve-hour day plus gasoline, and the railway seldom meanders more than two miles from the main highway, with trains averaging about 25 m.p.h. under perpetually clear skies — a winning combination, indeed! Mikado No. 52 crosses a deck girder bridge (*left*) twelve miles south of Al Gatrana. Legend has it that when the United States offered economic aid to the Jordanian government in the late 1950's, for "modern motive power," the Jordanians accepted the offer, but to the dismay of the U.S. AID officials, steam locomotives were specified! Embarrassed Americans then had to turn to the Japanese to build the engines, spending U.S. tax dollars there, instead of with American manufacturers who had phased out steam pro-

duction nearly a decade earlier. The result was a series of handsome Pacifics delivered in 1959, of which No. 81 (*below*), nearing Jurf ed Darāwīsh, is an example. At the bottom is a typical view from the highway as a 2-8-2 crosses the barren sands of central Jordan.

Lebanon

The former mainline connection of the Lebanese State Railways traverses the Mediterranean coast from Egypt, cuts northward through Israel, continues the entire length of Lebanon, then swings abruptly inland after crossing the Syrian border. Although the line carried light traffic and used some diesels, French-built steam power continued to handle an appreciable amount of the service north of Beirut well into the 1970's. Eastward from the capital ran the most interesting steam line to survive in Arab territory, in the form of a rack railway powered by big green and black 0-10-0T's.

Most of the industry and population of Lebanon extends from Beirut up along the coast to Tripoli, and it was here that several classes of standard-gauge 0-8-0's, supplied by French builders during the colonial period just after the turn of the century, worked in both freight and passenger service. On November 8, 1972, No. 104, a 1906 0-8-0, blasts out of the Beirut yards (*left*) with an oil train, then rolls right along the Mediterranean beach, passing local fishermen (*upper left*), as she brings her train of tank cars to the refinery at Nahr el Kalb. That night, No. 106 and 104 (*above*) repose in the Beirut engine terminal, and two days later, No. 34 (*right*), a smaller 0-8-0 with outside oscillating valve motion, built by St. Française in 1906, switches the Tripoli yard.

All photos this chapter, Ron Ziel

The Beirut-Damascus line, laid to a gauge of 42 inches, is a fascinating operation involving rack sections with grades up to 8 percent, a right-of-way that traverses the backyards of modern high-rise apartment houses and bleak desert, charming Victorian stations, snowsheds just twenty miles from the palm trees at its Beirut terminal, and ten-coupled green tank engines built by Société Suisse in 1924, as well as much more ancient power. While the coastline is semi-tropical, heavy winter snows cover the line as it reaches the 5,000-foot mark at Beidar Pass, just twenty-four miles from Beirut, necessitating the use of long tunnel-shaped concrete snowsheds to keep the line clear of twenty-foot drifts. As in Jordan, beautiful little masonry arch viaducts, such as the Jalala Bridge at Chtaura (*left*) abound, providing an exquisite setting for the 0-10-0T's, such as No. 301, shown here assaulting the western approach to Beidar Pass with a mixed train in tow. Sister 302 takes ample advantage of the Abt rack system as she struggles upgrade at Jamhouer (*right*) and Chouit-Aaraiya (*below*), her gears engaged in the rachet between the running rails. A much older Société Suisse product was 0-6-2T No. 6, built in 1894 and still going strong seventy-eight years later as the yard switcher (*lower left*) at Rayak.

Vietnam

Ever since the Crimean War of 1855, railways have been a decisive factor in every conflict of any size worth recording in the history books. At first glance, the Vietnam War may seem to have been an exception to this rule, but it was not. Early in the conflict the Vietcong, and later their North Vietnamese comrades, almost totally disabled the South Vietnamese rail system and followed up their initial successes with repeated attacks that lasted right through the war. It is common knowledge that the U.S. Air Force and Navy pilots who carried out the extended bombing campaign against North Vietnam, directed a large proportion of their attacks against the railway lines leading to China in an effort to stem the flow of supplies arriving from the Communist bloc. In all previous conflicts, the steam locomotive had been in the thick of the fighting and even in the twilight of its long and faithful service to man, the old puffers suffered their share of casualties on both sides in the Vietnam War. Indeed, a U.S.A.F. pilot with more than 200 missions in F-100's over North Vietnam told one of the authors that his favorite targets were North Vietnamese steam engines, which exploded into clouds of steam when hit.

Since the area now comprising Vietnam, Laos and Cambodia was a French colony during the age of railway building, the rail systems of what was then known as French Indochina were strongly influenced by the colonial builders. The locomotives were French-built and the numbering scheme was the same as that of the S.N.C.F. Engineering and track construction, as well as operations, are still carried out in the French manner, for the United States never sent railroad troops to serve in Vietnam as they had in World War II and Korea. Although American diesels had virtually displaced steam at the time of the January 27, 1973 cease-fire, the hapless French steamers had borne the brunt of battle. A dispatcher was more inclined to consider an old kettle such as Ten wheeler No. 230-333, her crew protected by an armor-plated cab as she takes a work train out of Hue on March 2, 1967 (*left*), much more expendable than a new diesel! One of the many acts of Vietcong sabotage occurred when a mine hurled Pacific No. 231-545 (*below*) ten yards from the track near Dieu Tri. The explosion and wreck, followed by a small-arms attack, killed nine people, including the engine crew. On April 13, 1967, another 4-6-2, No. 231-A001, still lies on her side south of Ba Ren (*right*), reportedly there since the French war more than a decade earlier. All of the steel plating had been stripped from engine and tender as they lay in Vietcong-controlled territory.

All photos this chapter,   Paul S. Stephanus

Probably the most famous section of railroad in all of Indochina was the Dalat rack line which managed to function, more or less secure, throughout the war. Bearing a striking resemblance to the Société Suisse 0-10-0T rack engines of Lebanon and built by the same Swiss company in 1923, 0-8-0T No. 40-303 labors hard and slowly up the 12 percent grade (*left*) allowing the photographer ample opportunity to jump off the train, run up ahead, take pictures, and reboard the cars. While soldiers assigned to guard the train patrol the area, some of the passengers alight to exercise as the engine takes water (*upper right and below*) at Eo Gio station. The complicated machinery, two cylinders of which operated the adhesion wheels and the upper set, that powered the rack cog (*middle right*) was a virtual masterpiece of the industrial designer's art. Even in the carnage and tragedy of Vietnam, so bucolic a steam operation managed to thrive!

Thailand

Of the entire Asian shore and the southern island-nations bounding on the Indian and South Pacific Oceans, only Thailand has never been a Western colony. This circumstance has had a delightful effect on the motive power of the meter gauge Royal State Railways of Thailand, which has been free to purchase locomotives where and when it pleased, unbeholden to any European rulers. On this relatively small system, engines from the factories of England, Germany, France, Switzerland, Holland, Japan and the U.S.A. have all worked side by side. Although threatened in the 1970's by aggression from their Communist-dominated neighbors, the Thais still found time to maintain what may well be the cleanest and shiniest stable of motive power of any nation anywhere. In fact, at every engine terminal, a stiff competitive spirit exists to reach the apex of excellence in maintenance and appearance. According to international steam expert A. E. Durrant: "Even Swindon could not have made a finer job of a *Castle* for a royal special, than did the regular driver of No. 840" — of the Thai Railway.

Called *MacArthurs* when they were built during World War II, the sixty-eight 2-8-2's (Nos. 380–447) of American construction were mostly veterans of the U.S. Army operation on India's Bengal & Assam Railway. Just after the war, the contingent arrived in Siam. As late as 1970, sixty-six were still in service, including No. 394, accelerating a passenger train past a temple near Bangkok's Thonburi Station.

All photos, Mike Eagleson, November-December 1972 (except as noted)

One of the American Baldwin-built 2-8-2's (*above*) visits Thung Song Junction. Virtually all of R.S.R.'s locomotives burn wood, a circumstance that is prompting the government to save more than 100 of the most modern engines for national emergencies after dieselization. Steam engine fuel is harvested in the nearby forests and sometimes hauled to the sawmills by little 0-2-2-0T Pachyderms (*left*) for cutting. Two men throw the cordwood to a woman on the tender (*lower left*) at Thung Song Junction, the destination of the elephant-hauled fuel. Finally, the wood is turned into steam-generating b.t.u.'s as the fireman of 2-8-2 No. 914 (*below*), now feeds the yawning maw of the Mikado's firebox. The newest of twenty-six three-cylinder 4-6-2's built by Baldwin in the late 1920's, No. 251 (*upper right*) works a local up the Kan Tang branch. Another spotless example of the World War II G.I. 2-8-2, No. 391 (*lower right*), an American Locomotive Company graduate, leaves the yard at Chong Khad.

Below, Paul S. Stephanus

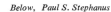

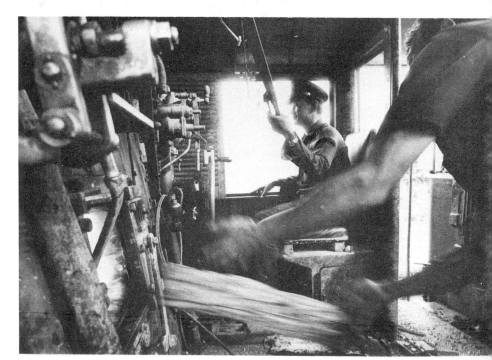

Leaving Ronphibun, Mikado No. 938 (*above*) is one of seventy built for the Japan Association of Railway Industry in 1949–50 as World War II reparations to Thailand. One of nine trim Japanese 4-6-2's built in the midst of the war, No. 288 (*below*) brings her mixed train into Ti Wang. Another reparations class consisted of the thirty Pacifics Nos. 821–850, including No. 828 (*right*) at Chong Khad.

Perhaps the most notorious of all railroad ventures was the Burma-Siam Railway, built by slave labor in World War II. An estimated 80,000 to 300,000 victims perished as a result of brutal conditions imposed by the Imperial Japanese Army. The most famous project of the "Death Railway" was the building of the bridge across the River Kwai Yai. Permanent rebuildings of the bridge during and after the war have long since replaced the original temporary timber structure erected by prisoners from the British Empire. The present span, now traversed by four daily mixed trains pulled by elegant little green Moguls, such as No. 730

(*above*), is a substantial masonry and steel affair. This photo was taken from the embankment of the original wooden bridge. Having crossed the bridge, sister No. 731 (*left*) leaves Kanchanaburi en route back to Bangkok. Forty-six of these Japanese C 56 class 2-6-0's were brought to Thailand by the occupiers during World War II. Streaking through the rice paddies, war reparations Mikado No. 939 (*below*) typifies the stunning character of the steam locomotives of the Royal State Railways of Thailand which, hopefully, will remain active through the mid-1970's.

Cambodia

Even as the era of the steam locomotives draws rapidly to a finale in virtually all of the nations of the world, and official dates of retirement are set in the rest, this most useful and abused devising of man continues to be exposed to the rigors and challenges of its youth — covering itself with glory to the very end. The beleagured southeast Asian nation of Cambodia is a striking example. As in Vietnam, the rail system was built by the French when they ruled Indochina. Naturally, the locomotives were French in character and in their numbering scheme. In the most recent conflict, the old steam locomotives carried the brunt of combat area assignments and suffered the most casualties. Brand-new diesels switched relatively safe and secure areas or remained stored in the sheds of Phnom Penh. In addition to their ruggedness and easier field maintenance, the old steamers had been paid for many years ago, making them expendable. The shining new diesels, however, were zealously shielded until the arrival of the elusive postwar era. In 1972 and '73, the dieselizing Thai Government Railways shopped a number of surplus steamers and sent them to Cambodia, adding more laurels to the record of steam power in wartime over a 120-year period. Pacific No. 231-509 (*left*) brings supplies to the army base at Tuol Leap while civilians carry their burdens (*above*) on their backs and in carts. The Phnom Penh shops hold a variety of power, including No. 231–502 (*below*, *left*), while outside, tank engine No. 131T-003 (*below*, *right*) awaits the heavy repairs that will enable her to return to the firing line.

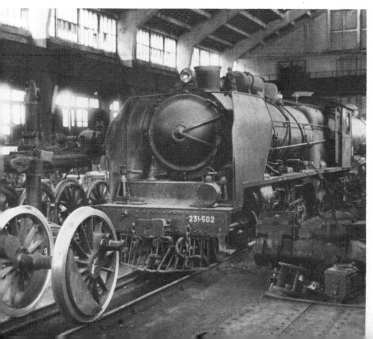

Four photos, Paul S. Stephanus, July 1972

207

Burma

Burma boasted an interesting variety of motive power which continued right into the 1970's, with two and a half times as many steamers than diesels in service in 1971. A Pacific (*above*) leaves the Rangoon yard and heads into the surrounding jungle with a long-distance passenger accommodation. Another straight-lined Belpaire-boilered 4-6-2 (*below*) runs light through the yard to couple onto her train. Storming past palm trees, a 2-6-4T (*upper right*) brings in a local passenger train. On the ready tracks, Baltic tank engine No. 764 (*right*) has taken water and a girl showers in the run-off from the water spout. Awaiting assignments to local passenger moves, a pair of 2-6-4T's (*lower right*) are serviced in the noonday sun. U.S.-built World War II 2-8-2 MacArthurs handled most of the heavy freight chores right up until the end of the steam era.

All photos, Paul S. Stephanus

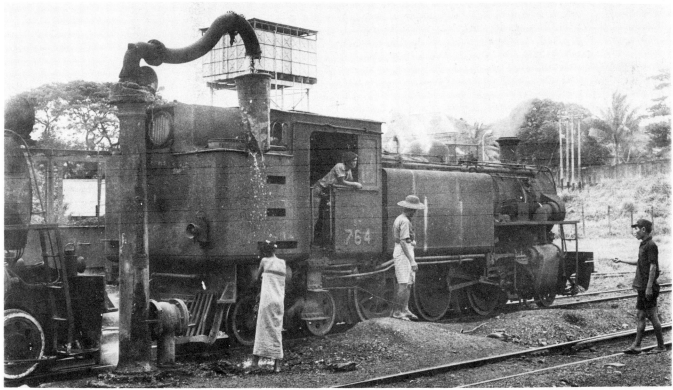

Indonesia

The steam operations of the multi-island nation of Indonesia were centered on Java and comprised every conceivable size and type of locomotive, from diminutive 0-4-0T enclosed tram engines up to 2-8-8-0 Mallets. The cleanliness and shining appearance of the *Perushaan Negara Kereta Api's* (Indonesian State Railways) steam power would make even a Portuguese or a Thai engineman envious, with spotless locomotives puffing along jungle branch-lines and city streets right into the mid-1970's. On the other hand, the operation of the P.N.K.A. often was chaotic, with inaccurate timetables and inept station staffs who delighted in quoting wrong departure times. The Indonesian Railway must be, according to the British *Continental Railway Journal*, "one of the most useless, inefficient systems in the world."

The usefulness and efficiency of the Indonesian State Railways, like many others in the world, may be questioned, but the beauty of its superbly maintained locomotives was beyond reproach. One hundred Mikados were delivered to the P.N.K.A. in 1951, including No. D52 083 (*upper left*) delightfully decorated with a pair of Indonesian stars, two coats of arms, a spread-winged bird above corporal's stripes and, in gleaming brass relief, an anteater whose tail is curled behind the smokebox door grab iron! This is a typical mixed train at Bandjong station. The Dutch supplied splendid locomotives for their colonies, and Pacific No. C53 17 (*lower left*) built in 1921, was no exception as she worked her last days on local freights out of Surabaja. Displaced by diesels on mainline passenger runs, all of C53 17's sisters had been withdrawn, but Djakarta-bound diesel trains now take two hours longer to make their runs! A classic 4-4-0, No. B53 06 (*above*), built by Hartmann in 1912, held down a mixed run on the line between Slahung and Madiun. Awaiting calls at the Bodjonegoro depot in East Java, the two immaculately maintained 4-6-0's (*below*) were constructed by Beyer Peacock in 1913.

All photos this chapter, L. G. Marshall, January, 1972.

To those accustomed to the homely little tank locomotives found on many railways, the handsome specimens of Java were a pleasant surprise. Most of the fifty-eight C28 4-6-4T's survived into the mid-1970's, pulling local passenger trains. No. 37 of the series (*left*), displaying two shining brass kangaroos beneath her headlight, worked out of Balapan depot in Solokarta. Massive for the 3′ 6″ gauge tracks it traverses, 2-12-2T No. F10 08 (*below*) switches the yard at Kertosono in eastern Java. Reminiscent of the Portuguese narrow gauge engines at Porto, 0-4-4-2T No. BB10 12 (*bottom*), one of the last of her class, switches at Ambarawa, near Semarang. At the big end of the spectrum, three classes of Mallets accounted for much line work and gained international fame as being the largest roster of compound Mallets to see the last years of world steam. A 2-6-6-0T (*upper right*) and a massive 2-8-8-0 (*lower right*) operate from Tijabatu, fondly known to railway enthusiasts as the "all-Mallet depot." Sporting a pair of brass elephants, 2-6-6-0 No. CC50 30 (*center right*) moves her mixed train out of Tasikmalaja.

The variety of big and little engines to survive so late in the diesel age on the P.N.K.A. was remarkable. The B22 class of 0-4-2T suburban tram engines was built at the turn of the century, and a handful, such as the one at Purwsari depot, Solokarta (*upper left*), could still be found. Forty-one little C11 2-6-0T's were built between 1879 and 1891. Incredibly, almost the entire class survived as shunters and even hauled some passenger trains eighty to ninety years later! No. C11 25 (*above*) was the station pilot at Rangkasbetung. In countries where electrification for trolley cars in cities was deemed too impractical or expensive, little enclosed tram locomotives handled trains right on the city streets. Surabaja eventually "modernized" with an electric tramway, but it was torn up years ago, leaving the field once again to such ancient dwarf lokies as the two (*left*) at Wonokromo depot. Slightly newer in design, a B17 class tram engine (*below*) runs through the streets in Tjukir. Rangkasbetung (*upper right*) was a stronghold for such exotic light power as C27-class 4-6-4T's and B51 4-4-0's. All of the B50 class, fourteen of which were built by Sharp Stewart in the early 1880's, were believed to be still operating nearly a century later! No. B50 11 (*lower right*) works at Madiun, where most of the class were based.

Taiwan

The Taiwan Railway Administration operates two principal and disconnected lines: the 2′ 6″ gauge on the east coast, between Hua-Lien-Chiang and Tai-Tung, which was virtually 100 per cent dieselized by the early 1970's and the heavily trafficked 3′ 6″ double track west coast main, which was still about 25 per cent steam-powered in early 1973. With 163 steam locomotives remaining on the active roster, a steady increase in both passenger and freight services and the railway making a profit on both, pressures for dieselization are not strong. However, plans for electrifying the mainline are proceeding, with completion scheduled for 1978, and that will retire the last steam locomotives. Railway construction on Formosa began in 1887, but virtually all of the building was done during the Japanese occupation that lasted from 1894 to 1945, resulting in a strong Japanese influence on the entire system, including the locomotives, all but eight of which were made in Japan.

Very Japanese in appearance, with long, graceful boiler, combination sandbox and steam dome and tall boxpok driving wheels, 4-6-2 No. CT274 (*left*) stands at the head of a northbound local at the main station in the Taiwan capital of Taipei on December 30, 1972. Several T.R.A. Moguls (lower *left*) are in various stages of repair at the main shops in Taipei. Tackling the grade on the double track main between Ying-ko and Taoyuan, Pacific No. CT272 (*right*) heads a local southbound from Taipei. Heading downgrade in the opposite direction, Mikado No. DT670 (*below*) represents the heaviest class of Formosan freight power as she coasts by with a northbound freight.

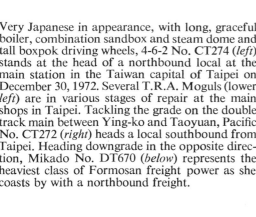

All photos this chapter, William D. Middleton

With forty-two cars of mixed freight in tow, 2-8-2 No. DT671 (*left*) takes the southbound mainline track at Shanchia on January 5, 1973. The remaining Formosan steam operation of great interest is the Ali Shan Forestry Railway, a 2′ 6″ gauge operation administered by the Taiwan Province Forestry Bureau, which connects with the 42-inch mainline at Chia-I. In addition to its timber haulage, the line carries many tourists up to the mountain resorts at Ali Shan. In just twenty-six miles, the railroad crosses 114 bridges, traverses forty-nine tunnels, makes several complete circles over itself and, almost at the summit, ascends four switchbacks. Grades are as steep as 6 per cent, accounting for at least fifteen Shays remaining in service, although diesel-hydraulics and railcars have taken over the mainline workings. The little two-cylinder, two-truck geared locomotives handle chores on the two logging branches beyond Ali Shan, as well as shunting and logging assignments along the mainline. Taiwan, with its variety of mainline power and the Shays, including No. 18 (*above and right*) at Ali Shan, had been often remarked in railway enthusiast circles, and in 1972, members of the Puffing Billy Preservation Society of Melbourne, Australia, purchased and moved Shay No. 14, built in 1912, to their own museum trackage Down Under.

Malaysia

By December 1972, diesels had almost completely enveloped the motive power ranks of the *Keretapi Tanah Melayu* — the Malaysian Railways — leaving just twenty active Pacifics. A little over two years previously, before large-scale diesel deliveries, half the roster of 200 locomotives was steam, including two sub-classes of 4-6-4T's and a number of U.S. Army 2-8-2 MacArthurs from the Bengal & Assam operation. Perhaps the most striking trait of the K.T.M. was the fact that as much as 85 per cent of its steam power utilized the highly sophisticated poppet valve steam distribution system. Another oddity lay in the extensive use of 4-6-2 Pacifics, sixty-seven of which were equipped with three cylinders and poppet valves, working as dual-service machines.

The last daily steam train from Singapore to the main junction at Gemas arrives at Segamat after dark (*above*) behind a 4-6-2, No. 563.10, named *Changi*. All of the trim Pacifics carried brass name plates on their boiler flanks, with Roman lettering on the left side and Jawi script on the right. Working hard out of Labis (*right*) No. 564.34 *Pekan* pulls a freight toward Singapore. *Following spread*: No. 564.36 *Temerloh* leaves Rengam.

All photos this chapter, Mike Eagleson, December 1972

By the winter of 1972–73 K.T.M.'s active steam roster was down to just a score of engines — all of them poppet valve three-cylinder Pacifics, the most complicated machines on the property. On most other railways the laws of limited maintenance would have condemned these intricate locomotives with the first diesel delivery. Although centered headlights were standard practice for all classes, a few engines, such as No. 563.11 *Langkawi*, had lamps reposi-tioned high on the smokebox (*upper left*) during the Communist insurgency of the 1960's, when gondola cars were pushed ahead of the engines to detonate mines. Trim Pacifics *Malacca* and *Changi* doublehead (*lower left*) from Batu Anam. The former engine works an oil drag (*above*) toward Singapore at Tenang. *Malacca's* complicated valve gear (*below*) improved steam distribution and operating efficiency, but greatly increased maintenance.

Korea

The railways of the Republic of Korea had hardly begun to rebuild after World War II when the invasion from the north, in June 1950, precipitated an even greater disaster. By the end of that conflict, three years later, a score of Chesapeake & Ohio switchers and a lone Erie light Pacific, No. 2524, had been sent to South Korea as gifts from the U.S.A. But by 1973, all of the American engines were gone. Only ninety-five steamers remained on the roster, ninety-three Japanese-built 2-8-2's of standard and 30-inch gauge and two native-built standard gauge 2-6-2T's. Their forced retirement was scheduled to take place within the year, due to several electrification projects which would release a number of diesels to usurp the steamers' hard-earned positions.

Although Baldwin in appearance, the Mikados which rolled out the last steam miles on the standard gauge lines of the Korean National Railways were all built by the occupying Japanese during the late 1930's, as were the narrow gauge 2-8-2's. Having held Korea for exactly a half-century following its annexation in 1895, the Imperial Japanese Army designed, built and equipped the K.N.R. When Nipponese domination ended, American influence took hold — including dieselization with 336 units, almost all of which came from General Motors. The U.S. Army rebuilt the K.N.R. after World War II and several times during the seesaw conflict of the early '50's. No. 318 (*left*) was a yard switcher at Subingo-Dong in Seoul on August 7, 1971. At Yongsan station, also in the capital, two Mikados (*above*) shunt in the snow the following February. Another 2-8-2 (*below*) works the yards at Suwon, on the main line from Seoul to Pusan, on August 13, 1972.

All photos this chapter, William D. Middleton

The last narrow gauge line to survive in the Republic of Korea is the 2′ 6″ gauge section between Nam Inchon and Suwon Suin. Operating seven diminutive Mikados, the line meanders for 125 kilometers through pastoral farmland and mountain hamlets, dealing mainly in mixed traffic trains and diesel railcars. Train No. 773 is shown with three different engines on three separate occasions in 1972: No. 13 (*above*) west of Suwon, No. 12 (below) at Song Do, and No. 14 (*lower right*) emerging from a tunnel near Suwon. On New Year's Day, 1972, train No. 774 (*upper right*) tackles the grade west of Suwon. Although all passenger service except the mixed trains on the line was dieselized by 1972, the remaining movements, including freight and switching, may well be the last South Korean steam operation, outliving the standard gauge Mikados.

Japan

In the land of the rising sun during the mid-1970's, the twilight of steam presented a paradox for the efficiency oriented, yet romantically sentimental soul of the Japanese. As the Nipponese nation advanced ever more rapidly along the course of ultra-modern economic development, seemingly obsolete institutions such as steam locomotives were cast aside. No other people, however, venerate the disappearing steam engines with more fervor than the Japanese. Indeed. the devotion of American and British enthusiasts pales into insignificance when compared to that of the Japanese masses.

Almost everywhere in Japan, huge color posters of steam locomotives (referred to simply as "SL" by the Japanese) adorn the windows and interiors of shops and public buildings. Postcard views and outstanding quality four-color process books depicting steam on the Japanese National Railways are widely circulated. Even in department stores, demure housewives carry shopping bags adorned with illustrations of J.N.R. steam power and such English phraseology as: "Hokkaido, home of the SL" or "C 62-2, Japan's greatest steam locomotive."

Rolling past the waning sunset in the twilight of Nipponese steam, a D 51 Mikado crosses a fill at Shikanotani on the island of Hokkaido.

All photos, this chapter, Mike Eagleson, November 1972

Japan is one of the very few countries where a rail-fan is not regarded as a freak. Indeed, since Commodore Perry brought the first steam locomotive to Nippon in 1853, the country has evolved into virtually a nation of railway enthusiasts. Scores of Japanese photographers carrying tripod-mounted Nikons and Pentaxes and plodding through ten-foot drifts on snowshoes, with bells jangling on knapsacks to ward off bears, were a common sight on the north island of Hokkaido as they pursued the last of the species SL in its winter habitat. As in Germany, steam had dwindled to fewer than 900 active engines by 1973 and total elimination was predicted within two years.

Also, in the manner of the *Deutsche Bundesbahn*, the J.N.R itself embarked on a well-conceived plan of preservation that included the Umekoji Museum in Kyoto — a large, modern roundhouse full of newly rebuilt locomotives of major classes destined to power excursion trains well into the distant future.

Meanwhile, the 2-8-2's, 2-8-4's, 4-6-2's, 2-6-2T's, 2-6-4T's, 2-6-0's and 2-8-0's were running out their last miles to the accompanying crescendo of thousands of clicking cameras in the hands of determined native and enraptured Western enthusiasts, as well as photographers simply in search of an animated subject offering infinite possibilities to express their art.

The world's first 2-8-2 was built by Baldwin of Philadelphia in the 1890's on order for Japan. The name *Mikado* — Japanese for emperor — was given to the new locomotive type. It is altogether fitting, therefore, that the dominant classes of J.N.R. steam in the last years were of the 2-8-2 wheel arrangement, with 45 per cent of all active steam in 1972 being represented by 434 D 51 and ten of the larger D 52 Mikados, the heaviest surviving freight power. Hokkaido-based D 51's work at Kuriyama (*above*) and at Otaru (*upper right*), and one of the earlier pre-World War II specimens, sporting a partial skyline casing (*lower right*), is at Shikanotani. One of only ten D 52's to outlast their 275 sisters into 1972, a 1943 veteran, No. D 52 235 (*below*), awaits a duty call at Oshamambe.

As in most countries, the mainstay of express passenger service in Japan was the 4-6-2 Pacific type locomotive. The first one in Japan was delivered by the American Locomotive Company in 1911. Of sixty-two C 55 class Pacifics introduced in 1935, just three remained active thirty-seven years later, including C 55 30 (*left*) at the headend of a Sōya Mainline train about to depart Ashahigawa station. Down to sixty-four of the original 201 built, two examples of heavier C 57 class 4-6-2's work local accommodations at Shibun (*right*) and Shikanotani (*below*) in 1972.

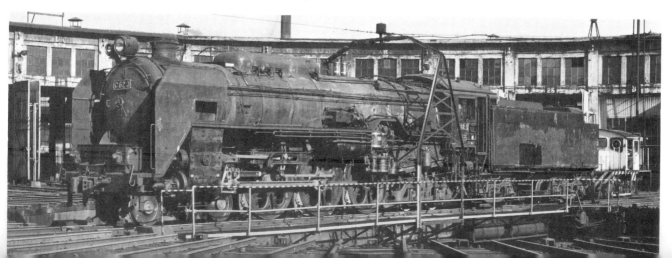

Handsome as tank engines go, C 11 2-6-4T's, such as the one pulling a local freight out of Moriyoshi on the main island of Honshu (*above*), once totaled 381. More than 300 were gone by 1972. Smaller but newer, 2-6-2T No. C 12 64 (*upper left*) switches at Otaru. The second most numerous class built (784) and the second in numbers surviving (181) into the final steam years, Consolidations such as No. 79616 (*far left*) departing from Kyogoku with a night freight dated back to 1913. The most beloved and handsome of all J.N.R. power were the forty-nine C 62 Hudsons rebuilt from D 52 Mikados in 1948. The last one in regular service,

No. C 62 3 (*below left*), rides the turntable at Otaru. With the forty-seven C 60 and thirty-three C 61 4-6-4's added to the C 62 total, J.N.R. possessed 129 Hudsons — second only to the "Great Steel Fleet" of the New York Central, which originated this classic type. A Mogul (*center left*) traverses the shore of the Japan Sea between Noshiro and Goshogawara on Honshu. One of a half-dozen D 61 Berkshires built in 1959 as Japan's final venture in steam, heads a mixed freight (*below*) from Rumoi through the Hokkaido landscape of early winter.

Australia

The country-continent of Australia is vast in distances and sparse in population, and the railways are concentrated mainly along the east and southeast coasts and in the southwest. Built as separate enterprises, with no overall coordination, Australian railroads wound up with their lines laid to five different gauges, and it was not until 1971 that passengers could cross the country without changing cars. The South Australian Railways were dieselized by 1969, followed rapidly by the Queensland Railways, the Victorian Railways and the Western Australian Government Railways, leaving just the New South Wales Government Railways with a few 4-8-4+4-8-4 Garratts and old 2-8-0 shunters to close out the age of steam "Down Under" in early 1973.

Total dieselization had become a peculiar trait of all the English-speaking countries. It would appear that the abolition of the English language might have preserved mainline steam operation — at least until the French joined the hitherto exclusively Anglican-tongued clique of major nations without steam power.

All photos this chapter, Mike Eagleson, April 1970

Known as "Big Boy" Garratts because their wheel arrangement of 4-8-4+4-8-4 made them the largest Australian locomotives, just as Union Pacific's 4-8-8-4 Big Boys were the heaviest in the United States, the forty-two standard gauge AD 60's, in the manner of the U.P. engines, were among the very last to go. AD 60 No. 6010 (*left*) leaves Mayfield, New South Wales. Shunting the Darling Harbour Goods Yard in Sydney, No. 1952 (*above*), a Z-19 0-6-0 with more than nine decades behind her tank, outlived many a class designed long after she had reach obsolescence. C 30 tank engine No. 3134 (*lower left*), built during World War I, passes Fassifern on the N.S.W.G.R. main. Rebuilt from some of the C 30 4-6-4T's in 1928–33, C 30T 4-6-0's such as No. 3088 (*bottom*) at Broadmeadow, still worked among their tenderless sisters forty years later.

The most popular of all railway photographic sites in Australia in the latter steam years was Fassifern Bank, north of Sydney. On Friday evenings, a score or more of native railway photographers would ride out from Sydney and environs to camp overnight alongside the steep grade where a constant parade of heavy coal trains — often double-headed — marched past. With tape recorders at the ready all night and cook fires burning before sunrise, the unshaven, bedraggled enthusiasts swapped steam stories and pictures until the light permitted action photography. Flushed out of the high weeds by a tandem composed of the first of twenty post-World War II D 59 class Baldwin Mikados, No. 5901, and a Garratt (*below*) a half-dozen Aussie enthusiasts get their "going away" shots on Fassifern. The most spectacular show on the hill was the double-heading of a pair of 4-8-4 Garratts (*right*) on an early morning heavy coal drag. Along with the "Big Boy" Garratts, survivors of the D 53 class of 190 2-8-0's built in the 1910's, were the last of steam to operate in regular service in Australia. Leaving Mayfield, No. 5353 (*lower near right*) storms past the tower while the signalman inside works his levers and keeps a wary eye on the photographer. No. 3801, one of five streamlined C 38 Pacifics reminiscent of the New York, New Haven & Hartford's Hudsons, rests in the Enfield Roundhouse (*lower far right*) in Sydney. This locomotive, the prototype of both the streamliners and the twenty non-streamlined 4-6-2's, has been preserved, as has No. 3813, the last of the green express passenger engines (*following spread*) riding the Broadmeadow turntable after bringing in the evening *Newcastle Flyer*. The preservation of these engines and multiple examples of every other important class on all of the major railways of Australia is indicative of the enlightened attitude of the railway managements. They all operate steam excursions, even to the extent of shopping retired locomotives on request for enthusiasts' specials — a magnificent policy which is diametrically opposed to that of the churlish and myopic managements of virtually all American railroads.

Gone were the maroon Garratts and the green Pacifics, and in the manner of limited maintenance which doomed most modern power at an early age but spared the ancient to turn the last miles in steam, the 3′ 6″ Queensland Railways operated Ten and Twelve wheelers during its final weeks.

4-6-0 No. 448 of the PB 15 class (*above*) leaves Redbank with a local passenger train. A 4-8-0 of class C 17, No. 705 (*below*), rides the turntable at the main shed at Ipswich in the sunset of both the day and the era of Queensland steam in Australia.

Months after the retirement of big 4-6-0's and modern R-class Hudsons such as No. 761 at Ararat (*below*), seventy of which were built in the early 1950's, Consolidations of J and K classes still shunted a few terminals and worked some of the 5′ 3″ branchlines during the Autumn wheat rush. The K class dated back to the early 1920's, but the J's were contemporaries of the Hudsons. Percolating on the turntable leads at Ballarat between switching assignments, J No. 507, K's Nos. 164 and 175 and J No. 516 (*above*) were under steam for the last time on Victorian Railways.

The State of New South Wales produces much of Australia's coal and at least two of its colliery lines were still in steam after the mainlines were finished. 2-8-0 No. 15 (*right*), built to Great Central Railway blueprints by the North British Locomotive Company for Great Britain's army in 1919, was one that subsequently went to J. & A. Brown Colliery at Hexham. At East Greta Junction, 2-8-2T's Nos. 19 and 10 (*below*) were among about fifteen of an original fleet of thirty-one of these big standard gauge tank engines operated by the South Maitland Railways.

The 5' 3" gauge South Australian Railways were the first, in 1969, to completely withdraw steam power. Strongly influenced by American design, the S.A.R. unabashedly copied the Pennsylvania's T-1 4-4-4-4's from which they developed their 520 series of named 4-8-4's in 1943. At least three of the twelve are being preserved, including *Essington Lewis*, No. 523 (*above*), but not *Malcolm McIntosh*, No. 522 (*right*), at Mile End Locomotive Depot. No. 526, *Duchess of Gloucester*, runs frequently in excursion service. Originally 4-8-2's, the ten engines numbered 500 to 509 were later rebuilt into semi-streamlined 4-8-4's. *Tom Barr-Smith*, No. 504 (*below*), shows definite Southern Pacific inspiration.

New Zealand

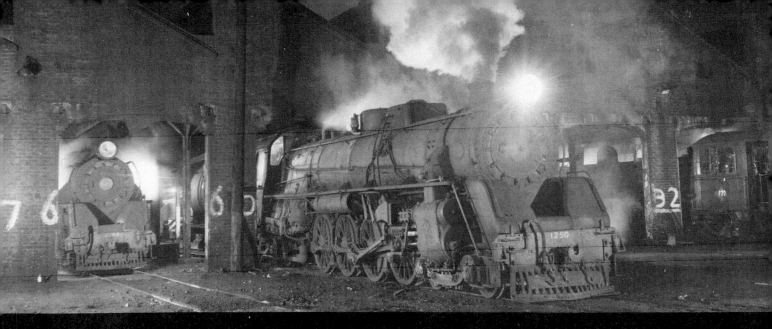

Blessed with a spectacular variety of natural beauty for so small a country, the New Zealand countryside often compelled even the most orthodox of steam photographers to concentrate on the setting in which the locomotives operated rather than on the engines themselves. By mid-1970, diesels had bumped steam completely off the North Island, and just two dozen of the New Zealand Railways' most renowned class — the J and Ja 4-8-2's — were running their last miles on the *South Island Express*.

Running from Christchurch to Invercargill, the *South Island Express* and other long-distance passenger services required two engine changes and several intermediate water stops, such as Palmerston (*left*) about midway along the line. 4-8-2's Nos. 1260 and 1250 (*above*) along with Pacific No. 777, in steam to supply the shed, offered some night action at Dunedin. Retired from regular service and awaiting disposition, 2-6-2 No. 847 and 4-6-2 No. 778 (*below*) lay cold and dead at Christchurch shed.

All photos this chapter, Mike Eagleson, May 1970

One of forty of the initial class J's built by the North British Locomotive Company in 1939, No. 1236 (*left*) works the northbound *South Island Express* along the 3′ 6″ gauge toward Christchurch. Storming up the Pacific coast at Katiki (*above*) a post-World War II Ja, No. 1250, built by the N.Z.R. at the Hillside Works in Dunedin, is one of thirty-five homemade 4-8-2's. The remaining sixteen, erected by North British in 1951, brought the total roster of J-class Mountains to ninety-one. The forty original J's, streamlined when built, were deshrouded by the early 1950's. A southbound Ja (*below*) traverses the beach where the sheep pastures meet the sea at Katiki. No. 1250, two 4-8-4's and three Pacifics, as well as a few smaller engines, have been saved for use on both islands as enthusiasts' specials, while a good selection of representative classes have been preserved by museums.

Colombia

While most Latin American railways are meter gauge or 5′ 6″ broad gauge, reflecting the Spanish influence in their design, a few are decidedly *Norté Americano*, with three-foot gauge and rosters of locomotives supplied mostly by Baldwin and the American Locomotive Companies. Such was Colombia, where yard-wide tracks meander among the foothills and pasturelands of the northernmost of the South American republics. Surprisingly modern, with all of its steam power built since the 1920's and most of it after World War II, the *Ferrocarriles Nacionales de Colombia* boasted massive 4-8-2's and stubby 1951-built Belgian Mikados. By the early 1970's, Colombian steam was virtually extinct.

All photos this chapter, Mike Eagleson, December, 1968–January, 1969.

The largest and most impressive of Colombian steam engines were the Baldwin and Porter Mountain types, erected in the 1944–1947 period. No. 106, a 1944 Baldwin (*left*), climbs upgrade into Cajicá. Pulling a mixed train, apparently the consist of virtually all Colombian movements, No. 110, resplendent in gaudy attire and with a 1960 Chevrolet chrome hubcap as a headlight decoration (*below*), marches away from the water plug at Nemocón with her *mixto* in tow. Less ostentatious, a younger Porter sister, No. 124 (*above*), rounds a curve at Sesquila.

The newest in years, if not in appearance, among the steam power of Colombia was a series of 2-8-2's built by Tubize of Belgium in 1951. The initial version — exemplified by No. 51 (*upper left*) and No. 52 double-heading with 4-8-2 No. 124 at Sesquila (*above*), both on the Sogamoso line — was rather handsome, with American-style front ends. The later Mikados, such as coal-fired No. 59, caught at night with 4-8-2 No. 75 (*left*) in Bogotá and leaving the capital (*right*) with a freight, suffered in appearance by comparison with their slightly older sisters.

Leaving the main station in Bogotá with the *primera* passenger train of the Colombian Railway, 4-8-2 No. 111 (*above*) proudly glides away while No. 127 struggles out of the freight yard with a Mikado assisting on the rear. As the earliest light appears on the eastern horizon, Baldwin 4-8-0 No. 81 (*left*) wheels her *mixto* out of Neiva. 4-8-2 No. 117 (*upper right*) impatiently flashes her oil fire and blows voluminous black smoke from her stack prior to departing Bogotá yard. Doubleheading a night freight from Colombia's capital city, 1928 Czechoslovakian Skoda-built Twelve wheeler No. 38, with the help of 4-8-2 No. 117, marches past a four-story signal tower (*right*). Today, all this is past, with German diesels having replaced steam in the 1970's.

In the early 1970's, Argentina was one of the most interesting of the world's countries in terms of numbers and variety of its steam power — ranking with China, India, South Africa, Turkey and Poland. It seemed that the reports of the earliest railway enthusiast trips almost served to alert the diesel salesmen, however, for no sooner had those who chronicle the antics of locomotives began packing their cameras and tape recorders than the government railway, *Ferrocarriles Argentinos*, announced a massive program of complete elimination of steam, using foreign and domestic-built diesels, by 1975. Fortunately, this goal was probably optimistic, perhaps by as much as several years. As late as 1969, close to 1,750 locomotives of at least seventeen different wheel arrangements and scores of sub-classes were still carried on the active rosters of the three main railway systems. It is inconceivable that the Argentinians — or anybody else — could afford an even higher level of myopic mania than did the railways of the United States in terms of the sheer waste caused by rapid dieselization.

When the various independent railroads were nationalized, they retained vestiges of their original identities in that each of the six systems was named after a great general in Argentine history. The three largest lines — the meter gauge General Belgrano, the 5′ 6″ gauge General Roca and the 4′ 8½″ gauge General Urquiza — carried on virtually all of the steam activity in the last years. Alone, the Belgrano line accounted for more than half of the active steam locomotives. Assaulting the bleak summit of the Andes at sunset, a Henschel 2-10-2, No. 1335 (*left*) heads toward Chile from San Antonio de los Cobres, on March 8, 1972, along the most spectacular line on the Belgrano. Among the last of steam to operate within Buenos Aires, a trio of 1906 North British 2-6-2T's (*lower left*) shunt the Roca's huge Constitution Station. One of the last survivors of twenty-eight Pacifics built by Maffei in 1910, No. 605 (*below*) works a train on the Belgrano suburban line from Buenos Aires to Gonzales Catan. By March 1972 only four of the 300 weekly passenger trains on this branch were pulled by the antique 4-6-2; the rest were all diesel.

All photos this chapter, Ron Ziel

The incredible variety of motive power remaining in Argentina as *la era de vapor* entered its final years, was too diversified to even begin to cover in the few pages allotted here, necessitating the omission of all but the most important classes. In 1948, Baldwin and the American Locomotive Company erected a class of sixty heavy 4-8-2 Mountain types for the Belgrano, including No. 1810 (*left*), her big headlight unfortunately replaced by a puny automobile lamp, shown leaving Recreo with a heavy freight drag. Henschel, Borsig and North British turned out 100 meter gauge wood-burning 2-8-2's in 1911, many of which had been converted to oil-burners in later times. No. 788, however, was among those still unconverted (*above*) as late as 1972, at Ledesma, on the northeastern line to Bolivia. Built by Alco in 1921, heavy Mikado No. 1471 (*lower left*) rolls northbound from Volcan, on the northwestern main, toward the Bolivian border. Lighter and older than the 1800's, the eighty-nine 4-8-2's built by Baldwin in 1921, and Henschel and Krupp in 1938, were the most numerous of the large Belgrano classes. Fresh from the shops, a Krupp Mountain in grey and black, with red, yellow and white trim, No. 881 (*lower right*) comes down the same line, south of Coya. *Overleaf:* After leaving the desolate station at Ingeniero Maury, named for the engineer who built the line from Salta, across the Andes to Chile, 2-10-2 No. 1342 climbs the 2 per cent grade with the weekly passenger accommodation to Socompa.

"Each canyon and stream that had known the hiss of steam,
Now echo to the diesel's roar . . ."

A Thousand Miles of Mountains (Northern Pacific Railway)

Known to the men of *Ferrocarriles Argentinos* simply as *Ramal C-14*, the line from Salta, which winds up into the Andes for some 250 miles to the Chilean border at Socompa, was begun in 1922 and finished in 1948. At that time the F.A. ordered fifteen heavy 2-10-2's from Skoda to work the line, along with the forty-six built by Baldwin and Henschel during the 1920's and '30's. Employing virtually every device developed by mountain engineering science, including tunnels, high bridges, two switchbacks, two complete loops and numerous horse-shoe and reverse curves, C-14, in the brief three years that it was known to railway photographers in steam, offered perhaps the most consistently spectacular operation in the Western Hemisphere. Until the diesels began arriving in mid-1972, the line had never been worked by any power other than the mountain-mauling Santa Fe 1300 class engines. Operating from terminals at Salta, San Antonio de los Cobres and Tolar Grande, crews worked a four-day swing across the arid, forbidding mountains, desert and salt flats for double pay. Those venturesome photographers who drove the unmarked and hazardous roads and rode the cars behind the laboring 2-10-2's were amply rewarded with some of the finest pictures of their careers. At Unquillal, in the early morning of March 16, 1972, No. 1334 pauses with the passenger train (*above*) as No. 1359 brings in a freight from Chile. In November 1971, No. 1355 (*left*) threads the El Alisal switch-back. The bridges of C-14 are awesome, but the most famous of all is *Viaducto El Polvorilla* (*below*) twenty kilometers beyond San Antonio de los Cobres. No. 1349 (*right*) is heading toward the great structure.

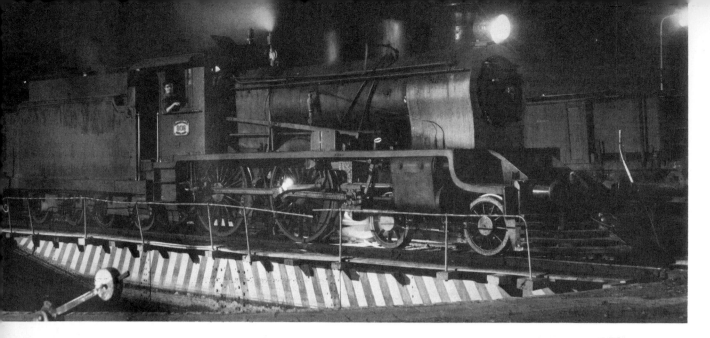

The General Urquiza, operating in the northeastern portion of Argentina, boasts some very rare locomotives indeed, including the last known standard gauge 4-4-0 (*upper left*) still in regular, mainline service in the Western Hemisphere. A diminutive 0-6-0 built by Neilson Reid in 1888, No. 31 poses alongside a road diesel at the Urquiza's steam, diesel and electric terminal (*left, center*) at Lynch, a Buenos Aires suburb, in her eighty-fourth year of active service. No. 512 (*lower left*) was one of six such 4-8-0 engines, built by Kerr Stuart during 1928–29. Running for nearly 200 miles across the arid plains and foothills of the Andes in southern Argentina, twenty-four Henschel and twelve Baldwin 2-8-2's, all built in 1922, power the mixed trains and freights of the little General Roca 2′ 6″ gauge branch to Esquel. A Henschel engine, No. 110 (*above*) nears El Maitén with her *mixto* while Baldwin No. 16 (*below*) climbs through a cut en route to Norquinco on March 3, 1973, handling the commerce of the towns of Patagonia.

Ecuador

The main railroad line in Ecuador is the Guayaquil & Quito, operating between the Guayaquil suburb and port of Duran and the capital of Quito, high in the Andes, and northward. Until 1970, this operation was virtually all steam and the exclusive preserve of Baldwin Moguls and Consolidations. When new Spanish diesels arrived, their first task was to take over the most spectacular portion of the line — a series of switchbacks on a mountain known as *Nariz del Diablo* (the Devil's Nose). A year later, little mainline steam activity was noted, although many engines were under steam and several were receiving major overhauls at the Duran shops. In times of heavy traffic, steam was out in force. For a quick stopover, just to sample Ecuadorian steam, Duran always was good for some activity, with at least five of the thirty-plus surviving engines switching the yard or awaiting mainline assignments. The Moguls, built in 1905, with most having been rebuilt and modernized in the 1940's, had once been painted a brilliant red with yellow trim, but by late 1971, only No. 11 still wore this livery. The other 2-6-0's, having been down-

graded to yard switchers, were painted the same black with white trim as were postwar 2-8-0's. Passing a monument erected by the local Lion's Club and lettered "homage to the railway," rebuilt (1944), Mogul No. 7 (*upper left*) shunts cars at the Duran shops. Little changed in her sixty-six years, unrebuilt No. 8 (*lower left*) switches a passenger train of wooden coaches into the decrepit Duran depot. In the adjoining shop building, identical cars were still being built of wood — completely from the railhead up and using only local materials and talent! 2-8-0 No. 53 (*above*),

outshopped by Baldwin-Lima-Hamilton in 1952, was one of the last commercially built American steam locomotives. Sister 54 and a slightly older 2-8-0, No. 47 (*below, left*) simmer in the Bucay shed. A worker with hammer and chisel (*below, right*) removes the builder's plates from one of the many derelict locomotive cadavers strewn around Duran, while a host of rubbernecks watch. The plates, removed for a fee of $6, were the heaviest souvenirs to be acquired by the author this trip.

All photos, Ron Ziel, December 1971

Chile

Politics and railroading are a fascinating study when their interests intersect, as is the case in many of the nationalized rail systems of the world. The steam locomotive has been an unfortunate victim of national vanities, for ever since the United States phased out steam, those nations that seek to emulate American practice have been destroying their steam locomotives, at times more out of political considerations rather than sensible economic thinking. Even the Soviet Union is known to have considered the rapidly dieselizing American scene before it, too, curtailed all further steam production. Happily for the steam enthusiast, if not for the citizens of Chile, the economic and social chaos brought on by the radical policies of Marxist President Salvador Allende Gossens in the early 1970's not only halted the dieselization process, but, with the rapid disappearance of Chile's foreign exchange, prevented the *Ferrocarriles del Estado* — the State Railway — from ordering more diesels. This resulted in scores of steam engines being taken out of scrap lines and sent through the shops for a reprieve of at least a few more years.

Unlike most Latin American steam operations, the broad gauge locomotives concentrated at the southern end of the Chilean State Railway are coal burners and the big steam terminals at Temuco and San Rosendo both have huge concrete coal tipples. Reminiscent of the coal docks on many U.S. roads, the structure at Temuco (*left*) fills the tenders of 4-6-0 No. 551 and 2-6-0 No. 580 on March 23, 1972. At the busy steam terminal at Victoria, 0-6-0T No. 463 (*above*) pulls a live but malfunctioning Ten wheeler, No. 548, onto the turntable that straddles a flooded pit. An affable, probably Baldwin built 4-6-0 of classic lines, with stack taller than her smokebox, No. 284 (*below*) shares a track with 4-8-2 No. 848 in the damp Temuco night.

All photos this chapter, Ron Ziel

The most common of all Chilean locomotives, accounting for almost a third of the nearly 400 active steamers in the early 1970's, is the class 57 2-6-0 Mogul, built by the German firm of Henschel in 1912. They almost completely dominate the many southern branchlines, handle local mainline passenger trains and short freights and do some yard work. In the traditional role of her class, 2-6-0 No. 522 (*upper left*) brings the morning Santa Fe-to-Los Angeles mixed train into the country way station at Candelaria, on March 22, 1972. Undoubtedly the biggest and certainly the best in Chilean steam are sixty-eight big 4-8-2's built by Baldwin in 1947 and Mitsubishi in 1952 and '53. They handle the heaviest mainline freights and fast passenger expresses. No. 852 (*upper right*) roars through Rihue at better than 50 m.p.h. In February, 1973, No. 819 (*left*) works hard past 2-8-2 No. 705 at Victoria and No. 862 (*right*) struts her freight out of Rihue and past the remnants of a Rumanian-built tractor demolished by a sister the previous day, while her progress is closely observed by a shaggy dog.

Some of the less common types of locomotives on the Chilean State Railway included such obviously British engines as 0-6-0 No. 394 (*upper left*) switching at Concepcion in March, 1972 — less than a year before the line from San Rosendo was electrified. Riding the turntable at Monte Aquila, 4-6-0 No. 546 (*left, center*) was one of twenty-five such engines built by North British in 1908. Usually used as switchers in the larger yards, these Ten wheelers still saw occasional service on local passenger trains. The standard heavy freight engine was the 70-class Mikado, which worked most of the mainline freights between San Rosendo and Porto Montt, along with the 800-series Mountains. Blasting southbound through Santa Fe, No. 714 (*right*) shows he worth, and No. 739 (*below*) takes a turn on the hand-operated table at San Rosendo, in scenes so reminiscent of railroading in the northeastern United States forty years earlier. In the far northern desert lands of Chile, where the barren Andes plunge right into the blue-green Pacific Ocean, the meter gauge lines were almost completely dieselized before the coming of Allende could have saved steam. And yet, as late as March 18, 1972, handsome little 2-8-2 No. 3555 (*lower left*) still switched the Northern Railway of Chile yard at Baquedano, the railroad center northeast of Antofagasta. Declining freight business numbered the days of the trim Mike and her distinctive Vanderbilt tender, for there were already enough diesels on the property of the F.C.N.C. to retire virtually all of the sisters of No. 3555 that still filled most of the stalls in the Baquedano roundhouse.

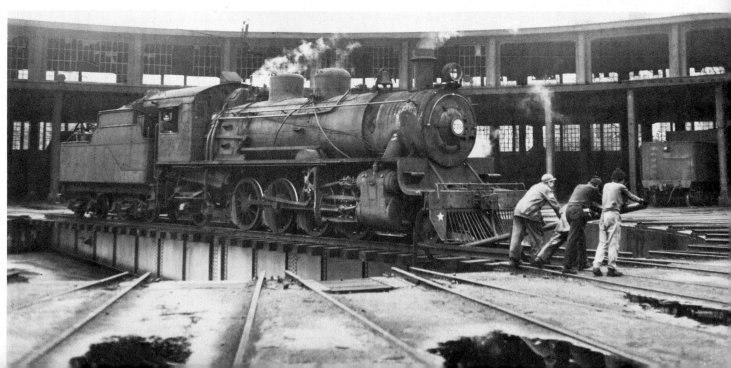

Brazil

As late as 1970, the trackless Amazon valley jungles, the coastal mountain ranges and the vast plantation lands of Brazil were still largely unchartered as far as types, numbers and locations of steam locomotives were concerned. Then, in a series of exploratory trips virtually unique in the annals of steam chronology, Californian Roy Christian and Briton Ken Mills began to ferret out incredible treasures in South America. Their most important strikes were in Brazil, a nation of disconnected railway lines meandering inland from the Atlantic coast in a variety of gauges, operating characteristics and eccentricities of motive power. While the comings and goings of virtually every engine in Europe was widely heralded for years, entire railway systems — 100 per cent steam powered — had been running in Brazil, their very existence completely unknown to the locomotive enthusiasts of the Northern Hemisphere.

Centered on São Joao del Rei, about 250 miles north of Rio de Janeiro, a thirty-inch gauge line runs for about 100 miles along the undulating banks of the Rio das Mortes, serving villages largely inaccessible by even the primitive dirt roads that branch off the paved highway. Boasting a roster of squat little outside-frame Baldwin 4-4-0's, 4-6-0's and 2-8-0's, some dating back to the 1890's and all in immaculate operating condition, the *Viacao Ferrovaria Centro Oeste* (V.F.C.O.) trafficked in light freight and a considerable passenger business. With a steady parade of mixed trains and freights, as well as work extras, the little engines put on quite a show. The most striking characteristic of the entire roster of homely little grungers is the absolute level of perfection attained in maintaining their appearance. Trimmed in red and silver, with gold lettering and boiler jackets agleam, Ten wheelers such as No. 39 (*left*) bringing her *mixto* toward Fazendo do Pombal near kilometer post 110, and Consolidations such as No. 68 (*below*) tooling out of São Joao before sunrise with brasswork shining and ornamental plastic propellers spinning, typified this eccentric operation. Even yard goats such as 4-4-0 No. 22 (*above*) sported silver smokestacks capped in glistening brass. These photos were made in March 1973, scarcely six months after the V.F.C.O. was first brought to the attention of the outside world by Roy Christian.

All photos this chapter, Ron Ziel

Another rare operation, whose existence has been known only since 1970, is a two-foot gauge industrial line owned by the Brazilian subsidiary of the Portland Cement Company. The line operates just beyond the northwestern limits of São Paulo, carrying limestone from a minehead through five miles of lush semi-tropical forestland to the huge cement plant at Perus. With about a score of engines on the roster, representing an incredible variety of types (2-6-2, 2-6-0, 2-8-0, 2-4-0, 2-4-2, 2-6-2T, 0-6-2T) in service, this is one of the last major steampowered industrial narrow gauge lines in operation in the world. The heavy power consists of 2-6-2T's with tenders added, such as No. 10 (*left*) working an ore train in November 1971 and No. 14 (*below*) whose tanks have been removed. These engines are identical to the famous U.S. Army "trench locomotives" which served on the front-line light railways of France in World War I. Barking up a tremendous storm as she assaults the grade with a trainload of empties from Perus, Mogul No. 7 (*above*) heads for the yard and the mine at Cajamar on February 29, 1972.

Among the more infamous railways of the world, because its construction and operation have taken such a high toll in human life, the *Estrada de Ferro*

Madiera-Mamore extends from the river town of Porto Velho, 1,500 miles from the mouth of the Amazon, through nearly impenetrable jungles, to the Bolivian border. Built around 1909 under such appalling conditions that the natives still call it "the railway of death" and say that there is a skull beneath each crosstie, the line may be abandoned due to the new trans-Amazonian highway system. Even into its last years, the E.F.M.M. ran trains only by daylight, for fear of Indian attacks, and the blue 1909 Baldwin Moguls and post-World War II Henschel engines were operated by the Brazilian army. To the amazement of visitors in 1971, a well-established museum (*left*) occupies the passenger station at Porto Velho, while 2-6-0 No. 11, named *10 de Julho*, switches the yard (*above*) close by the little white depot-turned-museum. More accessible and less precarious, the meter gauge *Teresa Cristina* division of the *Rêde Ferroviaria Federal S.A.* — the Brazilian Federal Railways — offers an alluring combination of heavy mainline operations and varied motive power, such as 2-10-4's on heavy coal drags (*below*) and trim, low-drivered 4-6-2's, such as No. 51 (*right*) shown pulling a derailed Mikado back onto the rails at Tubarão. Formerly owned by the R.F.F.S.A., a trim 4-6-0 and squat 2-8-0 (*upper right*) now switch the coal loaders of the Imbituba Dock Company.

The most enthralling steam operation in South America — perhaps in the entire world — to the American enthusiast just has to be the *Teresa Cristina*. The principal reason for the existence of this big-time mainline freight operation along the south coast of Brazil is to move solid trainloads of coal from various mine branches to the port of Imbituba, where it is transferred to barges destined for the industries of São Paulo. This heavy assignment is in the capable charge of a fleet of fourteen 2-10-4's, three 2-6-6-2's and a half dozen 2-8-2's, as well as a trio of 4-6-2's that handle some switching and work train assignments since the discontinuance of passenger service in the late 1960's. To further spice this potpourri in steam, several odd types, including a German 2-8-2 and a pair of handsome ex-R.F.F.S.A. 4-6-0's, a 2-8-0 and an 0-4-2T, work the docks, coal loaders and a power plant along the line. The most celebrated locomotives in Brazil, 2-10-4's Nos. 300–313,

built just before and just after World War II and believed to be the last Texas types operating in the world, bring their capacity trains down the mine branches to the classification yard at Tubarão. They then make their runs to Imbituba and return with empty wooden side-dump cars for the mine loaders. In March 1972, prototype No. 300 (*left*) climbs the grade at Imauri with thirty-three loaded cars, and sister 309 (*upper left*) tops 50 m.p.h. as she speeds a train of empties through Capivari on the mainline to Tubarão at sunset. Double-heading is not normally practiced on the *Teresa Cristina*, but in the clear morning air of March 15, 1973, Nos. 300 and 303 (*above*) combine forces to lift a long coal drag out of the yard at Pinheirnho — a spectacle reminiscent of the Bessemer & Lake Erie three decades earlier. Brazilians have no technical term for steam locomotives; they are known as *Maria Fumaca* — Smoking Marias — to railwaymen and laymen alike.

Since the axle-loadings of the 2-10-4's are too heavy for two of the four mine branches, the 2-6-6-2's built by Baldwin in 1950 and the Mikados, which are only slightly older, regularly hold down those runs. Clanking and wheezing through the jungles of the Rio Dezerto branch, No. 205 (*left*) works her loaded train between the minehead towns of Urussanga and Morro do Fumaca on March 2, 1972. A year later, sister No. 203 (*above*) digs in both sets of six-coupled driving wheels getting her train underway from the huge coal loader at Lauro Müller. At this time the third articulated, No. 204, occupied a bay in the spacious Tubarão shops where she was completely rebuilt — a job that took seven months. Leaving Tubarão at sunset with coal for the nearby power-generating station, postwar Baldwin Mike, No. 160 (*below*) sends her black plume high into the cobalt-blue skies of Santa Catarina province. Hopefully, this latter-day reincarnation of mainline super-railroading as it was known in the U.S.A. in steam days will last through the 1970's. Completely unknown beyond southern Brazil until its discovery in 1971, the *Teresa Cristina* has become the goal of increasing railfan forays. Diesels will not be welcome, according to the Chief Mechanical Officer, Gilberto C. Cabral, who is an avid fan of *Maria Fumaca* and frequently photographs his steel and steam charges. Often at the coal mines, a 2-6-6-2 will back her train beneath the loader, filling each car in turn and finishing with the tender. All diesel fuel would have to be brought in by ship, but the present power burns what it hauls, stating a strong case for its continued use; but then, so did the Norfolk & Western!

Guatemala

Ron Ziel

Built and operated by the United Fruit Company, the three-foot gauge International Railways of Central America were nationalized in 1970. The Guatemalans somehow wound up getting virtually all of the steam locomotives, while most of the diesels went to El Salvador. Although it once possessed 4-4-0's and 2-6-0's, by the time of nationalization the *Ferrocarril de Guatemala* (F.E.G.U.A.) was down to about eighty serviceable 2-8-0's and 2-8-2's and a solitary 4-6-0. A few years later, steam was being rapidly phased out.

By the end of 1971, virtually all steam road movements were handled by modern 2-8-2's built as late as 1948 by Krupp, Porter and Baldwin. A 1939 Krupp, No. 174 (*left*), crosses the Aqua Caliente Corozol Bridge between Guatemala City and Zacapa. Bound from Zacapa to the Caribbean Terminal of Puerto Barrios, a mixed train pulled by a 1947 Baldwin, No. 183 (*above*) leaves Gualan. The two classes of Consolidations, such as No. 110 (*below left*) a 1926 Baldwin at Zacapa and No. 68 (*below right*) a 1909 Baldwin at Puerto Barrios surrounded by the usual native yard crew and army sentry, were relegated to yard switching chores in the final years. *Following spread:* Outbound from Guatemala City an eastbound *mixto* crosses the 240-foot tall *Puente de Vacas* — the Bridge of Cows.

Two photos, Mike Eagleson

Ron Ziel

(*Overleaf, Mike Eagleson*)

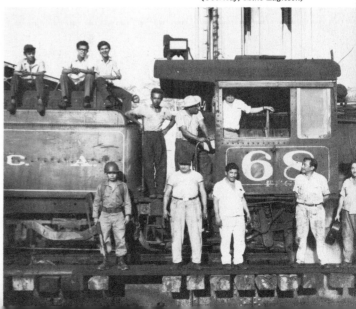

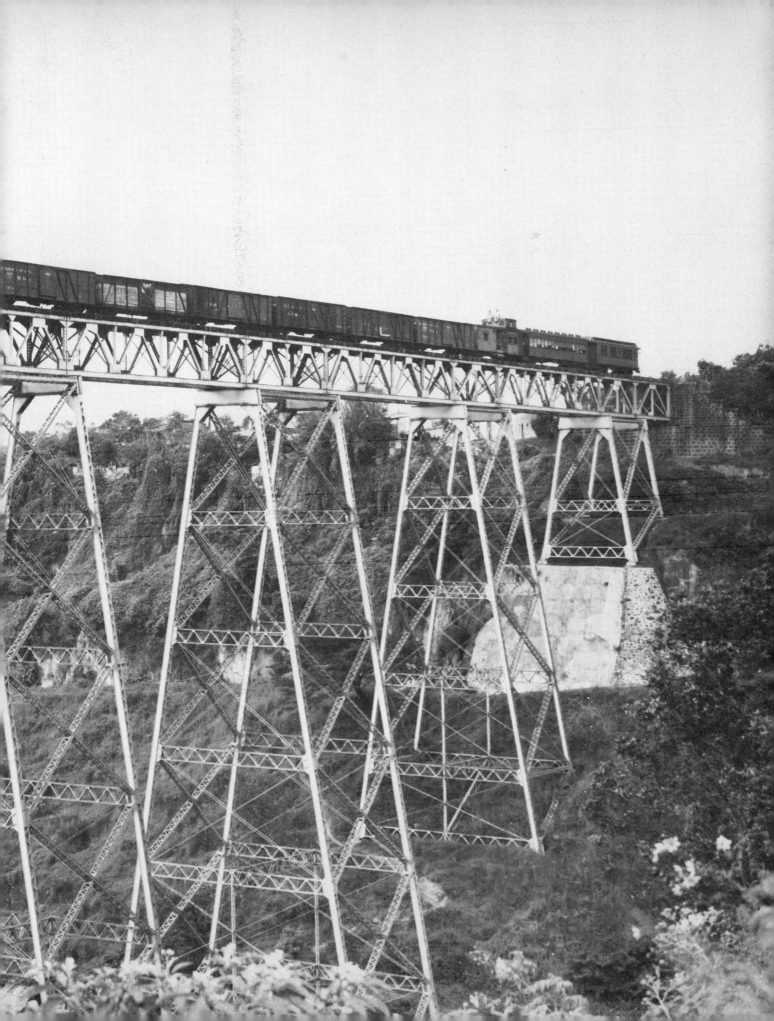

Bolivia

In just three years, from 1967 to 1970, the National Railways of Bolivia was almost completely dieselized. Only a few pockets of steam remained, including the last known Garratts to operate in South America. Among the survivors were several 2-8-4T's, including No. 554 (*above*) the switcher at Oruru. A unique feature of these squat meter gauge machines is the large fuel oil bunker built right into the cab roof. Apparantly, these engines are the last steam survivors of the old Antofagasta-Bolivia Railway, whose lines have been nationalized into the government systems of both Chile and Bolivia. Possessed of some truly modern power, including a dozen Japanese Mikados built by Hitachi in 1958, and heavy power, such as a brace of Baldwin 2-10-2's and a 4-8-4 — the latter only seen at night by rail enthusiasts and never photographed — the locomotive mysteries of Bolivia were disappearing before they could be accurately

chronicled. At the Argentine border town of Villazon, a thoroughly modern Japanese 2-8-2, resplendent in livery of black and red (*below*) switches freight cars just delivered by *Ferrocarriles Argentinos*. Perhaps the most renowned locomotive in Bolivia — certainly the most visible — was the yard switcher in the capital city of LaPaz. The engine, a two-truck Lima Shay, No. 508, was kept busy in both the passenger terminal (*lower right*) and on the freight sidings, some of which attained gradients of 7 per cent, probably ensuring a promising future for the old "sidewinder." While the Shay was taking water (*upper right*) the *maquinista* spotted the photographer and hastened to straighten the brass number plate. As the engine chugged about the yard, the plate invariably went askew, as is evident in the other photo. During the thirty-five minute photo "sitting," the engineer straightened the errant plate a half dozen times!

All photos, Ron Ziel, December 1971

Peru

In the 1960's, the Peruvian Corporation borrowed $18.5 million, most of which was used to dieselize the Central and the Southern railroads. A few years later, the British-owned firm defaulted on the loan and the Peruvian government took over the railways. In February 1973, the government announced plans to float a $40-million loan to modernize the rail lines, including dieselizing the largest of the narrow gauge operations — as if scrapping the well-maintained steamers, some of which were built in the 1950's, would improve service and increase revenues any more than the debacle of the previous decade accomplished! Meanwhile, the 36-inch gauge Huancayo-Huancavelica continued to operate its Baldwin Moguls, Hunslet Consolidations and Henschel Mikados in trim form. A 2-8-0 and two 2-6-0's (*above*) simmer outside the Huancayo engine shed. No. 107 (*below*), a trim green 1936 Hunslet product, brings the daily passenger train into Viques, en route to Huancavelica. Believed to be the oldest locomotive in the world still in regular service, 4-4-0 No. 2, built by Rogers in 1870, and No. 8, a relatively young Baldwin engine of 1910 vintage (*lower right*), sit cold in the Puerto Eten shed of the *Transportes Y Embarques del Norte*, accompanied by two 0-4-2's and a pair of 0-4-2T's, awaiting the arrival of a ship to unload. Just a few miles away, Mogul No. 6 (Baldwin, 1925) of the *Cia. Ferrocarril Y Muelle de Pimentel* switches trainloads of sugar along the half-mile-long dock (*upper right*) to be loaded into Russian ships. Swimmers and sunbathers at the adjoining beach pay scant attention to the green and red three-foot gauge engine. Happily, neither of these independent dockside operations will be affected by the government program.

All photos, Ron Ziel, February 1973

Mexico

Two photos, Mike Eagleson

Ron Ziel

Long after the last fires were dropped on mainline power north of the Rio Grande in 1960, steam continued to operate in impressive numbers and two gauges on the *Ferrocarriles Nacionales de Mexico*. To the young American enthusiasts who had vague childhood recollections of steam-stuffed roundhouses and Pennsy K-4's at 70 m.p.h., the first encounter with the Valle de Mexico terminal at Mexico City was an overwhelming experience. Big-time U.S.-built locomotives (including many second-hand from class I American lines) handled 60 per cent of the northbound freights as late as 1964. By 1966, the last of the 4-8-0's, 2-8-0's and 2-8-2's, as well as ex-Florida East Coast 4-8-2's such as No. 3316 (*below*) at Tula, were gone, leaving just a few of the big, modern, 4-8-4's, built in 1946, to handle local freights. Niagara No. 3047 (*left*) blasts upgrade out of Túnel Barrientos, north of Mexico City, past the most popular photo-spot in all Mexico for railway photographers. Outlasting the last four Niagaras by just a few months in 1968, the three-foot gauge lines east of the capital were dieselized or standard-gauged. On a happier day, March 21, 1966, narrow gauge 2-8-0 No. 262 passed Mt. Popocatepetl, the highest peak in Mexico, near Chalco (*lower left*) with the afternoon Ozumba-Mexico City local.

Ron Ziel

Canada

Although mainline steam operations officially ceased in Canada in 1960, at the same time they ended in the United States, the withdrawal of steam in America had been gradual, with most railroads running their last locomotives in branch-line and in yard service. In the Dominion, however, mainline workings in large numbers continued right into the early spring of 1960 and then were abruptly ended in an unprecedented wholesale slaughter of steam. The final, massive influx of diesels combined with a business recession to prematurely retire ranks of big and small, old and new and passenger and freight power. Newly shopped Canadian Pacific 4-6-2's, with brand-new flues, shining boilers and white-striped tires stood cold at St. Luc Shops, and Canadian National 4-8-4's were dispatched to the scrappers almost overnight, leaving just three or four engines available for occasional fantrips, including No. 6218, a glorious U-2-g Northern that ran excursions for seven years in eastern Canada. In an uproar of steam and contrived pandemonium, 6218 (*above*) thunders through an epic Canadian blizzard north of Toronto on January 23, 1966, as she carries 600 enthusiasts through the Grand River Valley of Ontario at 60 m.p.h. By the end of 1967, the only remaining Canadian steam was a smattering of lumber operations in British Columbia, a few colliery engines in Nova Scotia, and in Bienfait, on the wind-blown, snow-swept Saskatchewan prairies, two ex-C.P.R. eight-coupled switchers — a 2-8-0 and an 0-8-0 — moved coal trains to the interchange of their original owner. On a bitter cold Thanksgiving weekend in 1967, V-4-a No. 6947 (*upper right*) chuffs feebly about her switching chores. The finest in Canadian Pacific steam, an H-1-c Royal Hudson, whose future is now in the hands of the authors, was brought from the Dominion of Canada to the Commonwealth of Pennsylvania. During her complete rebuilding in 1972–73, No. 2839 (*right*), leased to the Atlantic Central Steam Company Inc. of Pennsylvania, is devoid of her streamlined shrouding and the crowns of England that were authorized by King George VI, after his 1939 tour of Canada, to be carried only on this class of forty-five locomotives. Scheduled to be operating in mainline excursion service by the publication date of this book, the authors' treasure was hopefully destined to become not only the most celebrated of North American steam locomotives, but the sole operating example of the ultimate in passenger power — a 4-6-4 — and the only streamlined engine as well.

Two photos, Mike Eagleson

United States

Last and definitely least in terms of serious steam operations during the twilight of world steam, the United States of America somehow contrived, in her renown eccentric manner, to account for more than 1,500 locomotives in various stages of preservation. They include about 200 in operable condition belonging to museums, private individuals and groups; five major railroads and the oft-contemptible tourist-trap operations. Being the first to dieselize, in 1960, the United States set the style for the rest of the world to emulate, yet more than a decade later several U.S. shortlines remained in steam.

Coincidentally, the two major freight operations to survive into the late 1960's dropped their last fires within a fortnight of each other, in late summer of 1968, and just a few hundred miles apart. Magma Arizona No. 7 (*right*) crosses Queen Creek Trestle on February 28, 1967, a year and a half before 2-8-0 No. 5 made the last steam run to the Southern Pacific interchange at Magma Junction in the Arizona desert. In southern Colorado and northern New Mexico, the three-foot gauge lines of the Denver & Rio Grande Western, the last sizable American all-steam line, ran its last freights on August 28–30, 1968. The final revenue train from Antonito (*lower right*) crosses the arid prairie to an assault on 10,015-foot Cumbres Pass. In addition to the D.&R.G.W.-owned Silverton excursion line, the Cumbres & Toltec Scenic Railroad runs steam trains over sixty-five miles of the old mainline. One of the last shortlines in the old South to utilize steam was the Mississippian. A brace of affable little ex-Frisco 2-8-0's ran until April 1967, when No. 77 (*below*) was still barnstorming through the cotton country between Fulton and Amory, Mississippi.

Mike Eagleson

Four photos, Mike Eagleson

Until 1972, when the U.S. Army phased out its last railway training facility at Fort Eustis, Virginia, several examples of the World War II G.I. 2-8-0's that still survived in the hundreds in eastern Europe were on call for training and excursion duties. The most spectacular trip was run for the National Railway Historical Society on September 3, 1966, when Nos. 612, 611 and 606 (*upper left*) triple-headed more than 700 passengers. The twenty-three-mile Reader Railroad in Arkansas, which in its last decade of operation carried passengers as well as freight, performed a spectacular rebuild of an ungraceful ex-Army sister, No. 1702 (*left*), which remained in service until the line closed in January 1973. South Carolina's Rockton & Rion (*above*) and Alabama's Mobile & Gulf (*below*) were the last common-carrier steam operations in the deep South.

Two photos, Mike Eagleson

Occasional excursions were operated behind mainline engines of the Union Pacific and the Southern Railway, while privately owned super-power 4-8-4 locomotives and other types were also available. From 1968 to 1971, the High Iron Company resurrected the Steamtown Museum's Nickel Plate Road 2-8-4 Berkshire, No. 759, which the authors chased by air across Ohio when she pulled the *Golden Spike Centennial Limited*. On May 5, 1969, with a 1928 Ford Tri-Motor (*right*) pacing the *Limited*, the 1944 Lima Berkshire tops 80 m.p.h. across the Midwestern plains. The final industrial operation consisted of ten serviceable ex-Grand Trunk Western heavy 0-8-0's at the Northwestern Steel & Wire Company mill (*above*) in Sterling, Illinois. It is ironic, indeed, that the nation which at one time ran more than 66,000 steam engines in class-one service and created by far the biggest, most formidable and most powerful of all steam locomotives should see the final ton-miles of common-carrier freight handled by a solitary 0-4-0 tank engine. The humble and anemic little grunger (*below*), No. 5 of South Carolina's Edgmoor & Manetta, still wheezes her way down three miles of weed-clogged rusty rails in 1973.

Ron Ziel

Czechoslovakia Mike Eagleson

Germany Ron Ziel

A PSALM OF STEAM

(Revised Standard Version)

The steam locomotive is my treasure; I
shall not want diesels.
It maketh me to lie down in green pastures
with my camera: it leadeth me beside the still
water towers.
It restoreth my soul; it leadeth me along the dirt
roads to scrap yards, wretched food and poverty
for its preservation's sake.
Yea, though I walk through the terminal of the
shadow of diesel, I will fear no secret police;
for my photo permit is with me; thy side rods
and thy stack-talk they comfort me.
Thou preparest a turntable before me in the
presence of the diesel salesman: thou anointest
my head with cinders; my tender runneth over.
Surely low, three-quarter sunlight and thick smoke
will follow me all the days of my life: and I
will dwell in the roundhouse of the LORD forever.

Amen.

Thundering down on the final signals of the age of steam —
the era which saw the greatest advances of man — the
lights read "high green" for the machine that symbolized
and gave meaning to the term "Industrial Revolution."
Here, in the final Twilight of World Steam, a British Railways Merchant Navy class Pacific, No. 35027 *Port Line*, is
silhouetted against the waning light at Brockenhurst on
August 15, 1966. For the great steam queens of scores of
other nations the twilight is also at hand and for many, the
eternal night of their going has already arrived. The lights
have turned to red, then dimmed and are slowly going out.
In a few fleeting years, all will be complete darkness on the
hundreds of thousands of miles of track where so recently
the greatest devising of man's mechanical genius ruled for a
century and a half.

Mike Eagleson

Spanish National Railways 4-8-2 Mountain No. 241.2090 at Quejigal, bound from the Portuguese border to Salamanca on June 4, 1968, when steam still ruled the rails of the Spanish main.

Mike Eagleson